You Can't Eat GNP,

I can eat English Muffin Bread!

Happy 42nd!

You Can't Eat
GNP

*Economics As If
Ecology Mattered*

Eric A. Davidson

FOREWORD BY
GEORGE M. WOODWELL

A Merloyd Lawrence Book

PERSEUS PUBLISHING

Cambridge, Massachusetts

Many of the designations used by manufacturers and sellers to distinguish their products are claimed as trademarks. Where those designations appear in this book and Perseus Publishing was aware of a trademark claim, the designations have been printed in initial capital letters.

A CIP catalog record for this book is available from the Library of Congress.
ISBN: 0–7382–0487-0

Perseus Publishing is a member of the Perseus Books Group

Find us on the World Wide Web at http://www.perseuspublishing.com

Perseus Publishing books are available at special discounts for bulk purchases in the U.S. by corporations, institutions, and other organizations. For more information, please contact the Special Markets Department at the Perseus Books Group, 11 Cambridge Center, Cambridge, MA 02142, or call (617)252-5298.

Text design by Heather Hutchison
Set in 11.5 -point New Caledonia by the Perseus Books Group

EBC 02 03 04 10 9 8 7 6 5 4 3
First paperback printing, March 2001

Dedicated to Bryce Talbert Davidson
He and his generation will live in interesting times

Contents

Contents

Acknowledgments

I thank several friends and colleagues who have provided helpful comments on drafts of this book in various stages of preparation, including Josh Bishop, Chuck Davey, Ken Davidson, Linda Davidson, Winifred Davidson, Andrew Deutz, Ross Gelbspan, Dan Markewitz, Dan Nepstad, Kilaparti Ramakrishna, John Shordike, Martha Tarafa, George Woodwell, and John Woodwell. My colleagues at the Woods Hole Research Center and at the Instituto de Pesquisa Ambiental da Amazônia are constantly offering thought-provoking, stimulating, and challenging discussions of many of the topics covered in this book. Many of the seeds of my ideas have come from them. George Woodwell has been a constant source of encouragement, insight, and inspiration. Michael Ernst turned my crude drawings into polished, handsome graphics. Tom Stone made possible the use of the satellite images of Rio Capim. Finally, I thank Merloyd Lawrence, who recognized promise in an early draft of this book and who provided a clear vision of how to render it worthy of publication.

Foreword

Ecology and Economics: In the Public Interest

George M. Woodwell

The Peace Corps set the young Eric Davidson down in a far corner of Zaire in central Africa in a tropical village so heavily dependent on local resources for the necessities of life that money, if it existed at all in that society, was not essential. He had come in hours from a life in the comfortable cocoon of the largest and wealthiest and most highly technologically developed nation to what most in that society would have considered as "camping out in the tropics." But it was not a weekend trip. He was a part of a grand experiment initiated by the United States in more imaginative days, an attempt to bring the disparate parts of the world together. Suddenly he was in a new world in which the objective day by day was neither profit nor wealth, but mere subsistence using the resources immediately at hand.

Subsistence, mere subsistence, within the limits of local resources, has been the objective for most of the people of the world over all of time, despite the ebb and flow of empires, fortunes, wars, and the constant accumulation of

larger numbers of people and technology. It remains, even
in this time of rapid technological change and accumulating wealth, the essential and immediate objective of perhaps as many as four to five billion people who are poor in
a world of six billion. This fundamental purpose has not
changed despite a ten- or twelve-thousand year history of
experience with agriculture and the development of civilization. Survival is marginal for most of the world, and it is
tied to a close dependence on local resources.

The issue seems remote to most in the wealthy western
democracies but even there the wealth is recent and the
difference is small between the freedom provided by confidence in the availability of the essential requirements for
life and struggling for subsistence. It is small in history and
small at the moment. The issue comes home to us all in
discovering the wasting humans living in the streets of our
cities, on the very edge of our culture and the edge of survival. It came home to me recently in an interesting way as
I considered a worthless stock certificate I had tucked in a
drawer some years ago and forgotten. The stock had be-
longed to my father. It was the only stock he ever owned
and it came to him as an inheritance along with ownership
of a farm in southern Maine where his mother had grown
up. The farm had been in her family for two hundred and
fifty years or more but became dysfunctional, at least as a
farm, in the modern economy of the mid-twentieth century. The small savings from the farm had been in the local
bank in 1929 when the bank failed. When the bank was re-
constructed years later, some small recompense from the
money lost was made by issuing new stock in the new
bank. That stock split and gradually accumulated as the
bank was transformed over nearly sixty years into a branch

of the Bank of New England. But then this bank also failed in the general banking disaster of the Reagan era. The stock became worthless. The sum total of the savings of agriculture on 160 acres in southern Maine over nearly 300 years, never much, had come for the second time to zero. Subsistence on the land was little more than mere subsistence for most of the populace, even in the United States, as recently as the early part of the 20th century, when even that opportunity succumbed to changing economic conditions.

The economic and technological expansion that marked the end of subsistence agriculture in New England in the first half of the twentieth century also marked the transition to global influences on environment in the second half. The signal moment was the first test of a nuclear weapon near Alamogordo, New Mexico in 1945, but it was followed by nearly two decades of testing of such weapons in the atmosphere and clear evidence of cumulative global contamination of air and land and water and birds and fish and people with radioactive debris. Even as the seriousness of that global influence was recognized and further testing was finally driven underground by agreement among all but China and France in 1962, evidence had accumulated that there was a parallel global contamination with long-lived chlorinated hydrocarbons including agricultural poisons such as DDT. It took another decade for the US to ban most uses of DDT. By that time, however, there was a spectacular record accumulated by Charles David Keeling of Scripps Institution of Oceanography of another global change. It was the annual increase in carbon dioxide in the atmosphere attributable to the combustion of fossil fuels. By 1970 there was serious concern that the changes in the atmosphere were enough to change climate globally. Suddenly the environment had pushed itself onto the political stage and had started the inexorable process of setting limits on human activities, not merely locally as long before established in English law and elsewhere, but now, globally. Economics and ecology are inseparably wedded in a high stakes game of politics. The stakes are the human habitat and its future.

Eric Davidson has defined the details of this game in clear and simple terms for every citizen. He defines it at a critical moment in the evolution of civilization when there is an emergent need for a new sensitivity to human rights, the right to clean air, to pure water, to a place to live in safety with access to food that is clean and safe. He writes during a time when a surge of democratic capitalism seems to be sweeping the world in all those nations where civil rights have any vigor at all. He writes about the core principles of economics even as the chaos of Russia, Colombia, Venezuela, Indonesia, and a score of other nations in 2000 reminds us all that no natural law of economics, government, or human nature assures the evolution of a stable or even tolerable human society. The free market, basic human greed, and a nominal right to vote for one's political leaders do not individually or collectively make a government that works in defending essential public needs. Much, much more is required, including regulations that define and limit by common consent both the powers of government and the capitalistic economic system.

The central mission of government of defending the interests of each from the potential depredation of all and the interests of all from the intrusions of each is compounded daily by growth in the human population, by the expansion of the technology for exploiting the world, and by the continuous expansion of human aspirations and expectations. A world of six billion people is failing at the moment in protecting us all from the rapid erosion of the human habitat through an open-ended global warming, through accelerating biotic impoverishment, and through the spread of poverty and human misery. The failure is intrinsic in a major change in the world from empty to full, from larger than needed to support the human endeavor, to smaller than needed. Can we re-fit our swollen global human foot into the finite shoe that is the earthly biosphere? Davidson says, optimistically, that we can, but only if we stop denying the diagnosis and start beginning the treatment, using the tools of both ecology and economics to find resourceful solutions to the environmental challenges of today and tomorrow.

The re-fitting requires recognition of some new rules from ecology that set limits on political action and on economics. That transition is what this book is about: the necessity for coupling into our economic and governmental system some of the facts of science, the laws of ecology that increasingly limit choices in management of human affairs in a world where pressures on all resources are soaring. The time when the objectives of economics and government could appropriately be focused within those disciplines ignoring global limitations of resources has passed. Suddenly, the focus has shifted to ecology. The shift has come rapidly, with some urgency, as we recognize that we are dealing with the whole earth and the issues have a last chance element not previously recognized. Suddenly, economics and politics, for far too long laws unto themselves in a world large enough to accommodate experimentation and failure, have become essential tools for adjusting human activities to the limits of the earth. Suddenly the laws of ecology have economic and political consequences never before acknowledged. Cherished assumptions of economics and the equally cherished compromises that have lubricated politics and governments for all of time no longer work. Suddenly the implications of the high discount rates favored in business and often favored in economics become clear as a serious cause of irreversible biotic and economic impoverishment.

The shock-wave is a heavy hit, numbing, almost beyond imagination. A torrent of denial runs through the business world and has support in government and in academic economics where ancient habits change slowly, if at all. But the speed of the transitions forces us all to face the realities of this new world and to take on as our own a continuous responsibility in minding the store Eric Davidson is telling us, not how to do it or what to do, but what some of the tools are that we all need in thinking about the new world of finite limits, soaring growth, and unbridled expectations. While the best may in reality be equity and mere subsistence, the hopes are much, much more for all the world's billions. In carefully measured terms, Eric Davidson shows us how to make the best we have work well.

You Can't Eat GNP

1 🍃

Whence Comes Wealth?

Three Fallacies About
Economics Versus the Environment

When Money Was Worth
the Paper It Was Printed On

Brahimu was the elementary school principal who, for a
modest fee, gave me private lessons in French and Swahili
in the remote African village where I lived in Zaire (now
called the Democratic Republic of the Congo). One of the
few villagers who earned a salary, he was a frugal, hard-
working fellow who really did save money under his mat-
tress. On the day after Christmas 1980, Brahimu's money
became worthless. I watched as he stood helplessly in dis-
belief in front of the dilapidated town hall, clutching a fat
wad of colorful money, each bill bearing the image of the
Zairian president. That money could have been used the
previous day to buy things—food, clothing, batteries, bicy-
cle tires—but now it was nothing more than colored paper.
The town leader could not give him any solace. The gov-

ernment had not told the local officials anything more than what we had all heard on the radio. This central African government had just declared that its paper money was no longer valid. New bills printed on a different color paper would become the official currency of the nation on New Year's Day. Everyone had a week to exchange the old currency for new bills at a bank. If there had been a bank nearby, a week might have been ample time, and the news would not have been so bad. But for Brahimu and the other villagers in this remote part of Zaire, the nearest bank was at the regional capital, reached only after an expensive three-day train trip, and the train did not always run on its weekly schedule.

Most of the villagers were subsistence farmers who did not need cash to buy food. They harvested their food from their own fields, and they hunted game and gathered snails and insects from the forest. I, too, was better off than Brahimu. By sheer coincidence, I had recently bought a good supply of salt in town, and it served as well or better than any flimsy paper money when it came to making trades for food. I also discovered a valuable asset in the old *Newsweek* magazines that the Peace Corps had sent me, which I had already read cover to cover. In fact, unlike my colorful, invalid paper money, the paper that *Newsweek* was printed upon had considerable value. The village children needed printed paper for folding into protective and decorative covers for their school notebooks. There was no other source of newsprint or printed matter in the village, so I had a monopoly on this market. I ripped out pages and charged what I thought the market could bear, which was one page of *Newsweek* for a mango and three pages for an egg. Pages with big color photos were worth twice those with only print. I managed surprisingly well without money.

No one from my village ever made it to a bank to exchange their outdated currency. Brahimu and others who had attempted to save money under their mattresses lost their entire savings as the old currency became worthless. The new currency eventually made its way out to the village by February and everyone flocked around to look at its brilliant new colors and the familiar portrait of their dictator-president who had just invalidated their old money. No one in the village, including me, ever really understood why the government had essentially taken their money, although we heard it had something to do with catching people off guard who were smuggling large sums of cash to Europe for some sort of ill-gotten gain. As far as I could tell, it was the peasant farmers and rural school teachers who were caught off guard. I eventually got my next cost-of-living allowance from the Peace Corps to carry on my public health projects, but I continued to trade salt and *Newsweek* for food because I had learned that they were actually more effective bargaining chips than the cash that had few uses in the village. With so little cash changing hands, this village of 1,200 people contributed virtually nothing to the gross national product of Zaire, which may well explain why the government did not seem to care much about them.

When Money Adds Up to Something

When money changes hands, as it did when you bought this book, the gross national product (GNP, see notes) of a nation increases. Economists use the GNP, essentially the value of all the products and services created and traded for money in our economy, as a measure of our well-being. The more money we spend, according to the GNP gauge of affluence, the better off we are. The value of the book is measured not by what you get out of it intellectually, emotionally, or spiritually but by how much money you paid for it. Whether we buy books about economics or ecology or buy hula hoops or handbaskets, the economy won't go to hell as long as we keep spending money. I'll do my part by pumping my portion of the book sales back into the GNP to buy dinner (paying with cash or plastic, of course; no one here takes *Newsweek*).

Our complex society is based on this trading of goods and services for money, but we seldom think about where the goods come from in the first place or what the consequences are of consuming those goods. Most of us do not need to worry about tilling the fields, fishing the seas, logging the forests, or disposing of the garbage. Other people do those jobs, they get paid for it, and it all contributes to the GNP. We are free to purchase what we want to feed our intellectual and gastronomical desires, and we can also feel warm and fuzzy inside about our contributions to the nation's growing GNP.

GET A LIFE — PREFERABLY A BETTER LIFE

At the same time, on the same planet, my former Zairian village hosts still live so close to nature that practically every bite of food, every scrap of wood, and every drop of water that they consume comes from their own hard labor, which they must exert most of the waking day in order to extract these resources directly from the forests, fields, and streams. Not only do they return home from the fields bone-weary each evening, but the women must then gather wood and haul water a mile or more before they can start cooking supper. Store-bought items like sugar, salt, and soap are luxuries that are not part of everyday life. Romantic images of peasant farmers or indigenous peoples living close to and in harmony with nature belie the difficulties and hardships in their lives.

The modern peasant relies on nature so heavily not because it is a noble or romantically appealing thing to do. Rather, she has no other option, except perhaps to migrate to the city where her family would become urban peasants. Urban and rural peasants are equally poor, but the rural peasant sometimes has a better chance of extracting enough to eat from nature. She knows from word of mouth that a small fraction of the people in her country and many other people in other countries have much easier lives, and she would give up the hauling of water and wood in an instant if she could. The most common question I was asked by my Zairian hosts was whether it was true that even the unemployed get paid in America. Although the question revealed ignorance about how our complicated unemployment insurance system works, it showed more importantly that even the peasant farmers of remote central Africa were aware of the existence of a "better life" even for the unemployed.

We seem to get closer to this "better life" as we get further and further removed from nature. The cars, stoves, plumbing, and shopping malls that make our lives so much easier also leave us several steps removed from the proverbial earth, wind, and fire that the peasant farmer knows so well. The more sophisticated our society, the more cushy

> I look forward to the day when statisticians add up the national accounts to take account of the depreciation of the environment. When we learn to do this, we will discover that our gross national product has been deceiving us.
> **—Arthur Burns, former chairman of the U.S. Federal Reserve Board**

our lifestyles, and the higher our nation's GNP, the less we need to think about where the basic necessities come from. If only that were true! It is not.

We may not need to think about tilling the fields every day, but we had better not lose sight of the fact that our wealth and our comfort are derived from a combination of natural resources—soil, water, air, forests, oceans, mineral deposits, climate—and the skill and ingenuity with which we utilize and manage those resources. If we neglect or abuse those natural resources, we undermine our own prosperity.

Most mainstream economists recognize that equating well-being solely with GNP is overly simplistic. Nevertheless, after recognizing the imperfections of GNP as a gauge of wealth, prosperity, and well-being, most mainstream economists go ahead and keep using it anyway. Many also ignore the role of irreplaceable natural resources as the foundation of our economic prosperity.

We need not go back to a peasant farming subsistence to appreciate the necessity of the natural resources upon which we depend, nor should we ignore GNP and other measures of modern economic prosperity, but we need to stop and remind ourselves that our modern, high-tech economy is still based on the wealth of a natural resource base. Unfortunately, three fallacies of mainstream economics persist and have lulled many people into thinking that we are insulated from the responsibility of being prudent stewards of our natural resource endowment.

THREE FALLACIES OF THE CURRENT MAINSTREAM ECONOMIC AND TECHNOLOGICAL MODEL

Fallacy 1. Marie Antoinette Economics

The first fallacy comes from those economists who explain how the world works primarily in terms of markets and exchanges of money. In the debate about global warming (which is discussed in Chapter 5), one economist argued that we need not worry much about the effects of global warming on the economy, because the only sector of the economy that he considered strongly influenced by the climate is agriculture, which contributes only 3 percent of the United States' GNP. Like Marie Antoinette's suggestion that French peasants without bread could eat cake, this view of how the world works seems to suggest that if the crops fail, the people could eat the 97 percent of the GNP that remains. The importance of the natural resources upon which agriculture and most of our food supply depend extends far beyond their direct 3 percent contribution to the GNP. Food production is a good example of how imperfect GNP is as a gauge of our well-being, because we all recognize how important food is to our daily lives, even if agriculture is of only small direct importance to the GNP. The role of forests in providing clean water and habitat for plants and animals and in regulating the climate is another example of how GNP fails to count the true value of natural resources.

At present, the mainstream of the discipline of economics generally fails to recognize the extent to which human economic activity is dependent on the biological and physical condition of our environment. Until very recently, modern economics and ecology have been studied separately, with little constructive communication between disciplines. However, the human economic system and the biophysical ecological system of the earth are inextricably linked. The premise of this book is that the economic system will fail if the ecological system is not carefully managed. The inverse, which is also true, is that a failed economic system creates desperate people who will destroy the ecological system.

Fallacy 2. Custer's Folly

The second fallacy assumes that the technological cavalry will come over the hill in time to save us from ecological disaster. The cavalry usually arrives in the nick of time in Hollywood westerns, but it arrived too late in the real world of the life and death of General George Custer. Will technology replace the internal combustion engine in time to avert global warming caused by burning coal and oil? We have already changed the atmosphere so profoundly that the next few generations will have to cope with a warmer, less hospitable world. Technological developments in modern agriculture have averted massive famine, so far, as the human population has grown to 6 billion people. As the population climbs to 8 or 10 billion, however, and as soils erode away and groundwater is depleted and contaminated, will technological development keep pace and enable us to provide more and more food and clean water with fewer and fewer natural resources?

We will need the best that technology can provide, but to be more cautious than Custer, we had best not rely solely on future technological developments to clean up the messes that we are now making with our current bad habits. Prudence dictates that we slow population growth, prevent soil erosion, conserve groundwater, and stop polluting the atmosphere. Future generations will benefit from these essential natural resources under any scenario of technological development.

This second fallacy is based on the common assumption in economics that new technological developments can almost always find substitutes for a natural resource when that resource is depleted. Although this assumption is valid for some natural resources, it is not valid for the most important ones. It is true that when we run out of oil and coal, we will substitute solar energy or fusion or some other source of energy. Clever engineers may also come to the rescue by inventing technological substitutes if and when we run out of copper, cobalt, or any of a number of minerals that we use in industry now. The technologically optimistic economists, however, fail to see the difference between these substitutable natural resources and the resources that humans will continue to require (at least for the next several generations) and for which there are no substitutes. These essential natural resources include soil, air, fresh water, oceans, and forests.

The technology argument can be taken to absurdities that are not relevant to our lifetimes or to those of our children and grandchildren. Perhaps crops can be grown hydroponically (in nutrient solutions without soil); seawater can be desalinized and purified on a large scale; air can be filtered; the regulatory effect of forests on climate can be substituted with shields and mirrors placed in orbit above the earth. Although it would be foolish to rule out categorically any of these technological dreams, it would be much more foolish to assume that any of these extravagant ideas would be feasible on a large scale within the next few decades, if ever. That leaves us still needing soil, air, water, and forests for our livelihood. Technology can either help us or hurt us to make wise use of these resources, but it cannot completely substitute for them.

We also depend upon the generally favorable climate of the earth. We and the plants and animals that we depend upon for food and other resources are well adapted to the present climate, but human-caused global warming is underway that will change our social, economic, and ecological adaptations to climate. Unfortunately, as discussed in Chapter 5, several technological optimists are currently arguing that future technological advancements will allow us to cope with global warming, so that we need not do anything now to reduce emissions of the greenhouse gases that are causing a rapid change in climate. How can these modern optimists be any more certain about the timely arrival of their technological cavalry than Custer was about his cavalry? Would it not be more prudent to take steps now to mitigate and avert problems resulting from global warming and, at the same time, work on developing possible future technological solutions?

Fallacy 3. False Complacency from
Partial Success (or "Not Beating the
Wife As Much As Before")

The environment has progressed from a fringe issue in American politics in the sixties to a mainstream issue that pollsters and politicians take very seriously as we enter the twenty-first century. The Clean Water Act and the Clean Air Act of the 1970s have significantly improved the quality of surface water and air in the United States. Some polluted rivers and lakes that used to catch on fire are now clean enough for swimming. We have reduced poisonous lead emissions into the environment with the use of unleaded gasoline. A number of writers have pointed to these successes as evidence that environmental pollution is no longer a serious problem, but would you respect a man who says he no longer beats his wife (or Mother Nature) as often as he did before? A bit of progress is no reason for complacency in a world where forests are being converted to ranches, farms, and abandoned land at an astounding rate, where the genetic diversity of plants and animals is declining and species are going extinct at unprecedented speed, where fisheries are collapsing, where soil is eroding faster than it can be regenerated, where heat-trapping gases are accumulating in the atmosphere, and where groundwater is becoming depleted and contaminated.

Most of the progress that has taken place came after hard-fought battles. In the 1970s, for example, the oil industry vigorously opposed the phase-out of leaded gasoline, claiming that unleaded gasoline would cost too much, that it would adversely affect the GNP, and that the harmful effects of lead were still unproven. They were clearly wrong. Gasoline in the United States is now cheaper, when corrected for inflation, than it was when it was leaded. Taking the lead out of gasoline has had no discernable negative impact on the GNP or on our economic prosperity. The effects of lead poisoning on children's neurological development are now well documented. This victory for the environment and for our health required both the social and political will to change (rejecting complacency and recognizing an unacceptable environmental pollution problem) and a technological solution (development of affordable unleaded gas).

ECOLOGY, ECONOMICS, AND TECHNOLOGY IN PARTNERSHIP

The following chapters examine environmental deterioration of soils (Chapter 2), air (Chapter 5), water (Chapter 6), and forests (Chapter 8), showing how the three fallacies—narrow economic perspectives, the assumption that technology can replace essential natural resources, and unwarranted complacency from partial successes—have led to severe environmental problems. Rather than predicting environmental gloom and doom, however, I believe that these fallacious economic assumptions can be removed from economics and that economics and ecology can become mutually supportive. For each of these environmental problems, we already know enough, or nearly enough, to avoid or solve the problems, just as there was a solution to pollution from leaded gasoline. Knowing what needs to be done from an ecological perspective, however, is not enough. Proper economic analyses and policies are also needed to help provide incentives for designing and implementing those solutions. The combined skills of technologists, economists, and ecologists are needed.

A back-to-nature approach will not work—there are too many of us for that; and those who have been forced to live close to nature, like my Zairian peasant farmer hosts, live a very hard life. In searching for solutions, however, we must avoid the illusion that market economics, alone, makes the world go around, that technology will solve our environmental messes in time to avoid hardship, or that enough progress has already been made to afford complacency.

How humbling it is to realize that despite our sophisticated technologies and complex international economies, we still need to eat, drink, and breathe and that we depend on natural resources to provide the food, water, air, and climate to do so! Although this fact is no surprise to many readers, it is remarkably absent from editorial pages, news magazines, and talk shows. Instead, we hear more and more about the excesses of environmental protection—"ecoterrorism" Rush Limbaugh calls it—and the harms of environmental protection to our economy and to the GNP.

This common shortsighted view sees economics and ecology as incompatible, when, in fact, these two disciplines must join forces if humans are to continue to prosper. Several tools and habits of mainstream economics, such as cost-benefit analysis (Chapter 3), discounting (Chapter 4), and ignoring externalities (Chapter 5) are often misused, but each can be turned around to be a positive force for environmental management.

Several new thinkers in economics, who call themselves *ecological economists,* are trying to do just that. Like mainstream economics, this new form of economics relies on the profit motive of the marketplace. However, unlike mainstream economics, ecological economics attempts to define market values for the essential services provided by the environment that are not currently being traded effectively in the marketplace. As described in the following chapters, finding clever ways to inject the value of the environment into the marketplace creates economic incentives that reward and promote ecologically sound management practices and that help maintain a healthy environment. How do we place values on soil (Chapter 2), forests (Chapter 3), air (Chapter 5), groundwater (Chapter 6), and diversity of plants and animals (Chapter 8) when these are items that you can't buy in a store and that are hard to work into a cost-benefit analysis? Moreover, how do we assess the value of these essential resources for future generations? The tools of both ecology and economics are needed to address these questions. When these disciplines are combined, they show that a healthy environment yields food that you *can* eat as well as provides the basis for a prosperous economy.

2 🌿

Richland for Dirt Cheap

Two Views of the Value of Soil

RICHLAND CREEK DRAINS a forested area near Raleigh, North Carolina, where I first started learning about soils. "Richland" is a common name for creeks in this part of the world because the English settlers recognized that the low-lying lands near the creeks were often naturally fertile due to the sediments deposited during spring flooding. By the time that I walked that land some two hundred years after its naming, this once rich farmland was growing only spindly pine trees.

This tract of 500 acres was first purchased from the newly created state of North Carolina in 1779. Ownership changed hands five times during the following ninety years. Each of these owners used the land fairly lightly. By 1860, only 70 acres of this forested land had been cleared for agriculture, and most of that was used for growing what was called "Indian corn."

Mr. William H. Burroughs bought the land in 1869 for $3,500. He cleared another 130 acres and planted cotton, which was the beginning of the end of agriculture on this land. Mr. Burroughs was not alone as a cotton farmer. By

1880, 59,000 acres of cotton had been planted in the surrounding county. Only forty years later, however, this parcel of land along Richland Creek, like much of the other cotton farmland throughout the southeastern United States, had become so severely degraded that it was virtually useless for agriculture, and it was given back to the state of North Carolina.

What happened in the interim of only fifty years to convert rich land to land that was so degraded it was given away? In those days, cotton was planted in rows spaced far apart, so that a donkey and cart could be driven between the rows. With so much bare soil exposed between rows, the rich topsoil was quickly eroded away in this hilly landscape, leaving exposed a barren, unproductive subsoil that the farmers abandoned. Some farmers were known to have boasted about "wearing out" three farms in their lifetime. Although the soil started out naturally fertile and rich, the purchase price of land was cheap enough that a farmer could afford to wear out one farm and then use a small fraction of the cash from his cotton crops to buy another parcel for the next farm.

When the Division of Forestry at the State College took over the land in 1939, the foresters decided to plant loblolly pine at very close spacing, because they feared that only a small percentage of pine trees would survive in this highly degraded soil. Forty-three years later, when I was a graduate student in forestry, my class visited this planted forest of spindly pine trees. Less than one inch of new black topsoil could be seen on top of the abandoned eroded farm soil. In another forty years, perhaps there will be slightly more than an inch of topsoil. Long after I am retired, forestry students may be able to return to this same forest and learn the lesson of how slowly soil develops and how poorly even pine trees grow on degraded soil.

The cotton farmers of the late nineteenth and early twentieth centuries were reacting to the economic situation of their time. Land was plentiful and cheap, and cotton yielded good profits. Economists would argue that these farmers were acting rationally in terms of doing what provided them the greatest profit for their investments of capital and labor. Ecologists, on the other hand, see a sad, irrational legacy of abuse of the land by previous generations, which limits the potential use of the land today and for several generations to come. Clearly, ecologists and economists think differently about the value of soils and other natural resources. Both ecologists and economists have visions of the way the world works, but their visions seem worlds apart. Ultimately, however, these worldviews can be and must be reconciled.

THE ECOLOGIST'S VIEW

The ecologist views the flow of energy and matter through the natural environment as a pyramid. At the wide base of the pyramid is the soil (or the ocean for aquatic ecologists) from which plants get nutrients, water, and a foothold. Green plants depend on the soil, as well as sunlight and air. The herbivores (cows, deer, certain insects) that eat the green plants are less abundant than the plants, so the pyramid narrows as we move up the food chain. The carnivores (wolves, lions, spiders, humans) that eat the herbivores are less numerous still. Humans eat both meat and vegetables, and because humans are seldom eaten by other carnivores, we usually place ourselves at the top of the pyramid.

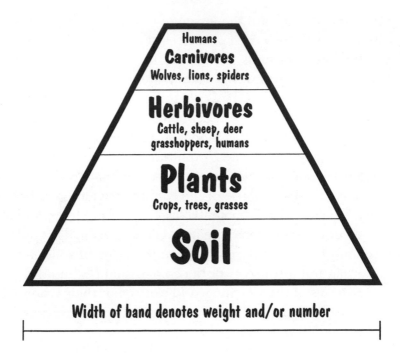

The Ecologist's Pyramid

The ecologist's pyramid shows how dependent we are on a stable resource as our base. If the soils erode away or become contaminated, it is obvious that part of the pyramid will collapse, affecting us as well as all of the other plants and animals in between. The width of each band in the pyramid is not determined by a judgment of which group of organisms is more "valuable" than another, but rather it is calculated from the amount of energy (calories of food) or mass (weight of organisms and their chemical makeup) produced and consumed at each level.

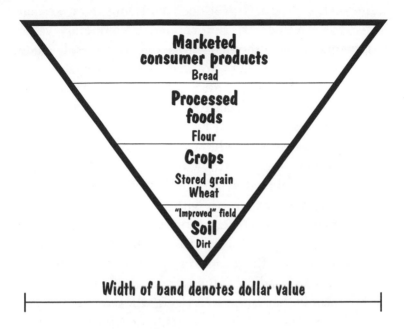

The Economist's Pyramid

THE ECONOMIST'S VIEW

Neoclassical economics is the mainstream modern form of economics that applies sophisticated mathematics to the ideas of *classical economists* such as Thomas Malthus, Adam Smith, David Ricardo, and John Stuart Mill. Neoclassical economics emphasizes how supply and demand affect markets. The neoclassical economist's pyramid is inverted. No wonder economists and ecologists have been slow to find a common vision! Each level of the neoclassical economist's pyramid is measured by the monetary value of the products produced and consumed at that level. These values are determined by monetary value judgments made in the marketplace, where people trade their money for products. Not many people will pay much for soil (which is the source of the expression, "dirt cheap"), so soil is a very narrow band at the bottom tip of the inverted pyramid. If a farmer *improves* the land by clearing the forest, tilling the soil, and growing a crop on it, it gains economic value. Ironically, the soil will lose some of its native fertility after a few years of tillage, making it less able to support a crop with the same input by the farmer, but the neoclassical economist still considers the soil *improved* because it is now a farm that has more commercial value than the *unimproved* woods or prairie. The farmer's work has added value, so the crop is a wider band in the pyramid than is the soil. Using wheat as an example, the value continues to increase as the crop is milled and processed, bread is baked and transported to market, and until it is ultimately served for consumption.

Similar examples would trace the fish in the sea as having little or no value in the economist's pyramid until they are caught, processed, packaged, transported to stores, bought, and served for eating. Similarly, minerals must be extracted from the ground and forests must be cut before economists assign them significant value.

The economist's pyramid appears to be precariously balanced on a tiny point that is the soil layer. If the soils were washed away, would the economist's pyramid come tumbling down? If the soil were eroded completely away and it could not be substituted by another technology, the neoclassical economist would agree that the upper layers of the pyramid would suffer. If the soil resource is only partly

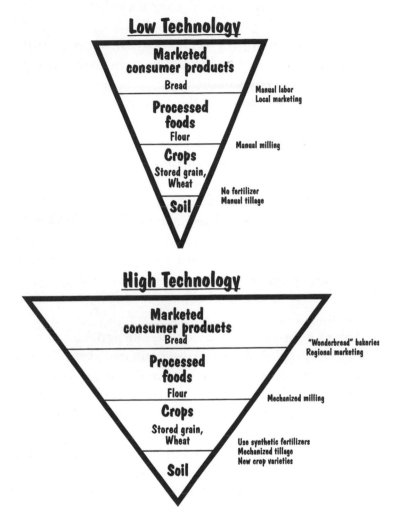

Economist's Pyramid: Low and High Technology

depleted and does not disappear entirely, however, the economist would argue that improved agricultural technologies (more fertilizer, more pesticides, irrigation, new crop varieties) could produce the same crop yield and the same value at higher levels in the pyramid while using less soil at the base. In effect, technology substitutes for the lost soil. The degree of sophistication of the technology determines the sharpness of the angle of the pyramid.

History provides some evidence in support of the neoclassical economist's view that technology can, to a limited extent, replace soil. Before the advent of chemical fertilizers, about five times as much land was needed to feed a family than is needed today. In addition to using chemical fertilizers, the breeding of new crop varieties, the use of pesticides, and irrigation have increased yields further. Of course, these technologies have deleterious side effects on rivers and groundwater, which are discussed in Chapter 6, but the fact remains that agricultural technology has allowed us to get more food value out of each unit of land and soil.

Had it not been for these advances in agricultural technology, the predictions of massive famine made by Thomas Robert Malthus, a classical economist of the late eighteenth and early nineteenth century, would have come true. Malthus predicted that human population would grow at an ever-increasing rate (exponentially), whereas the best that could be expected of the agricultural technology of the day was a gradual and constant (arithmetic) rate of growth in food production, and so a shortfall in food production relative to the needs of the human population would be inevitable. The population has indeed increased nearly exponentially, but growth in agricultural productivity has more or less kept up with the demands of an exploding population (although over 1 billion people are now malnourished, largely because of problems of food distribution and local economies). Maybe Malthus was both behind and ahead of his time. He did not foresee the power of technology to yield more food, but the jury is still out as to how far we can allow the soil base to become depleted while still feeding a rapidly growing population. I have great faith in technology, but it would be foolish to assume that the soil resource can be depleted indefinitely, and it is simply wrong to assume that the soil can be replaced completely for feeding humanity.

What Is Soil? How Fast Is It Made?

My only childhood memory of thinking about soil is that I would be in big trouble if I tracked it across the living room carpet. Growing up in a small city, soil was nothing more to me than dirt. Nearly half of the world's population is concentrated in urban areas, and most of those nearly 3 billion people probably have had similarly limited first-hand experiences with soil. The suburban weekend gardener may appreciate rich black soil, and most people probably recognize that their food comes primarily from farms, where tractors are commonly seen plowing the soil. It is probably fair to say, however, that thoughts of soil usually end with the gardening hobby, the distant image of farmers plowing their fields, and the dirt tracked into the house.

Soil is more than dirt. It is a complex mixture of minerals and organic matter that can provide a rich medium for abundant plant growth. It takes a long time, however, for Mother Nature to make good, fertile topsoil. Dead leaves and roots are gradually mixed with the clay minerals by the activity of worms, mites, bacteria, and other organisms living in the soil. This mixture has the right combination of nutrients, aeration, and water-holding properties to nour-

> Perhaps the most serious obstacle impeding the evolution of a
> land ethic is the fact that our educational and economic sys-
> tem is headed away from, rather than toward, an intense con-
> sciousness of land. Your true modern is separated from the
> land by many middlemen, and by innumerable physical gad-
> gets. He has no vital relation to it; to him it is the space be-
> tween cities on which crops grow.
>
> **—Aldo Leopold**

ish the plants. When the topsoil is eroded away, exposing
the deeper subsoil layers that have not developed this rich
mixture of organic matter with minerals, abundant plant
growth cannot be supported. Depending on the climate
and the type of vegetation, the formation of an inch of new
topsoil can require anywhere from fifty years to several
hundred years.

Where I now live on Cape Cod, Massachusetts, there is
a thriving market for topsoil produced from compost. Several small companies collect yard rakings, wood chips, and
other organic garbage to make giant compost piles. After it
is well composted, they mix this material with sandy subsoil and market it as *loam* to landscapers, gardeners, and
homeowners. In essence, the composting speeds up the
rate of forming new soil. I was able to create a small gar-
den in my yard where the builder of the house had scraped
away the topsoil long ago. At over $100 per truckload,
however, this technology is fine for the weekend gardener
who enjoys the hobby as much as the expensive tomatoes
that it yields, but this technology will not replace degraded
soils on millions of acres of farmland and forests.

The Extent of Soil Degradation

Although there are no global databases that tell us exactly how much land has experienced significant soil erosion, several soil scientists and agronomists have studied this question and have come up with some startling estimates. The area of the world that is already so extremely degraded that the soils can essentially never be reclaimed is almost as big as the entire state of Iowa. The area that has lost about half its agricultural productivity and therefore is of little utility today and would be extremely expensive to reclaim is larger than the area of the United States east of the Mississippi. The area with soil degradation severe enough to reduce agricultural productivity by 10–25 percent and that would be moderately expensive to reclaim is bigger than the entire United States, including Alaska. And the area that is lightly degraded, where declines in productivity of up to 10 percent have occurred, but could be fully recovered is yet nearly another United States-sized area. Hence, an area equivalent to twice the area of the United States has had significant soil degradation, and over half of the degradation is severe enough to require a lot of money to reverse. Still more soil degradation is irreversible on timescales relevant to human lives. Not surprisingly, some of the most degraded soils and largest areas affected occur in the most populous nations of the world—India and China—but soil degradation has also occurred in both industrialized and developing nations, both rich and poor nations, and where both modern and traditional farming practices are used.

Why has this soil degradation happened? Soil erosion, salinization, and waterlogging have been recognized problems, and their causes have been understood, perhaps since the beginning of agriculture itself. In the ancient civilizations of the Middle East, elaborate systems of irrigation were used that provided just enough water for crop growth, but not so much that toxic salts would accumulate in the soil (salinization). These systems worked for centuries, perhaps thousands of years, but a combination of wars, migrations of peoples and their cultures, population growth, and in some cases, the substitution of the wrong modern technology has resulted in the breakdown of these irrigation systems and the development of soils that are so full of toxic salts that they can no longer be used for agriculture.

The Central Valley of California contains soils that can be used very productively for agriculture when irrigated. In a large part of the region, however, the soils contain the natural element selenium, which leaches out into the irrigation water and must be disposed of somehow. Selenium is not harmful in the quantities found in these soils, but it accumulates to harmful levels if irrigation drainage water is allowed to evaporate, leaving selenium-concentrated salt behind.

Soil scientists in California in the late 1800s recognized this problem of salt accumulation and warned against irrigating these soils without proper disposal of the drainage water. But when irrigation water was provided by various government-funded projects after World War II, mostly in response to lobbying from the farming corporations that profit from taxpayer-subsidized irrigation, the only provision for disposal of the selenium-containing drainage water was a big evaporation pond in the center of the valley at Kesterson. There were political and economic reasons for this decision: (1) there was heavy opposition to discharging into San Francisco Bay the agricultural wastewater that might be contaminated with pesticides, fertilizers, and salts; and (2) it would have been too expensive to pump it over the coastal mountain range and discharge it directly into the Pacific Ocean, assuming that a politically feasible discharge point could have been found there. So the easy answer was to let the water evaporate away in big ponds constructed in California's Central Valley.

As predicted, the selenium became concentrated in the ponds, it poisoned nearby wildlife, and millions of dollars are being spent to clean up the mess. In the meantime, no long-term solution to the selenium disposal problem has been reached. Although the problem at Kesterson has drawn a lot of media attention, hundreds of "mini-Kesterson" evaporation ponds are accumulating toxic salts throughout the American West because of inappropriate irrigation management.

These errors in management in the arid American West are not very different from the mistakes that resulted in soil degradation in the arid agricultural lands of the Middle East. Many scientists knew of these ancient lessons, but their voices were not heard, and short-term profits carried the day. Modern Israeli and Arab soil scientists and agronomists have developed sophisticated irrigation systems that make very efficient use of scarce irrigation water and that avoid the problems of salinization, but these techniques have not been widely adopted in the equally dry American West, because cheap, subsidized irrigation water is still available, and the accumulation of toxic salts is being hidden in small evaporation ponds dispersed across the vast countryside.

Here is a case where many economists share the same point of view as environmental activists. Economists argue that if water projects were not subsidized by the government, then the marketplace would determine the price of water delivered to farmers. Irrigation water would then become more costly and would be used more wisely, and as a result, salinization of soils and selenium poisoning of wetlands would be reduced. The underlying values of soil and water are not included in this analysis, but the cost of *moving* the water to an arid agricultural area can be determined very well in the marketplace, and the true high cost of moving huge volumes of water long distances would provide the needed incentives to use both water and soils more wisely. Unfortunately, the farming corporations that benefit from the water subsidies use their profits to buy the political influence needed to maintain the subsidies, thereby preventing the marketplace from doing its work and resulting in depletion of soil and water resources.

Erosion is another predictable and preventable problem. The cotton farmers of North Carolina were by no means the first to have caused soil erosion. The ancient Greeks and Romans noted that the topsoil would erode away when their hilly forests were cleared for agriculture. Although they recognized the problem, their answer to a declining agricultural base at home was to expand their empires into other regions to obtain, among other things, more agricultural land. That approach worked for several centuries, but when their empires began to crumble (for many reasons) the previous loss of fertile soils near to home contributed to their weakness.

A few years ago, a government-sponsored resettlement program in Indonesia paid loggers to clear forests for agriculture, but the program managers failed to have the loggers trained so that the topsoil would be preserved. Instead, the topsoil was scraped away by bulldozers along with the logging debris. Not surprisingly, the newly resettled farmers could not grow much without topsoil, the settlement scheme failed, and the displaced people cleared more forest by hand in search of fertile land.

These are examples of both ancient and recent history of soil degradation, but what is the current rate at which further soil degradation is proceeding? One technological super-optimist, the late Julian Simon, cited statistics of declining rates of soil erosion in the United States as evidence that advances in agricultural technology can solve this problem too. He was correct that we have the technological know-how to prevent soil erosion, but his rosy presentation of the statistics is an excellent example of the third fallacy described in Chapter 1—false complacency from partial success (or not beating the wife as much as before). In fact, the annual rate of soil erosion in U.S. cropland is now averaging about 13 tons per hectare (5 tons per acre), which is at least ten times faster than the rate at which Mother Nature makes new soil. This rate of erosion is lower than it used to be, so soil conservation efforts have had some success, but we are still depleting the soil resource. Although Iowa is still a tremendously productive "bread basket" of the United States, the topsoil there is about half as deep as it was when the first European settlers broke apart the prairie sod with a plow. So far, we have substituted fertilizer for the native fertility of the lost soil, but the soil loss cannot be tolerated indefinitely because the soil cannot be substituted entirely.

That is the good news. The bad news is that average erosion rates are about ten times higher in much of the rest of the world compared to the United States, which makes erosion rates about one hundred times greater than the natural rate of formation of new topsoil. I have seen peasant farmers clearing forests and planting corn on steep slopes in Mexico and Zaire. They know that the soil will wash away and that they will have to abandon the land and clear more forest in a few years, but they have no choice. They need to feed their families, and their governments have offered little or no assistance in finding better farming practices. Although they do not have empires to exploit as did the ancient Greeks and Romans, the modern tropical peasant farmer can use the same basic strategy as that of the ancient farmers or of the American cotton farmers of the nineteenth century: When your cleared forest land becomes unproductive (due to erosion or to loss of fertility), go seek more forest to clear.

If we do not learn from the lessons of the past and find alternative agricultural practices, the result will be equally devastating. Just as the once-forested Mediterranean region is now nearly void of trees and has severely eroded soils, the tropical forests will also disappear, and where the slopes are steep, these tropical soils will also erode unless the current trend of rampant tropical deforestation is reversed.

The amount of land becoming so degraded that it can no longer be farmed is not known with great accuracy, but if the best estimates are correct, between 0.5 percent and 1 percent of the currently cultivated land of the world is being lost each year. If this trend is not stopped, 25 percent to 50 percent of the current agricultural land will have been lost by the time my six-year-old son is in his fifties.

Although new cropland is also created by cutting down forests, it is obvious that this rate of land degradation can only make the task of growing food and protecting forests all the more challenging in the future. The current rate of land degradation demands urgent action that requires the combined skills of ecologists, economists, and agricultural technologists.

The problem is not that we are dumb—we understand why the soil becomes degraded and how to avoid it—but that collectively we are forgetful, some of us are greedy, and some are desperate. And there is no shortage of inept managers and policymakers ready to condone and implement practices based on greed and desperation. That we have made progress in some places proves that we can find solutions, but it is no reason for complacency. The forces of greed, desperation, and ineptitude are strong adversaries that impede prudent application of knowledge and sound reasoning.

RECONCILING THE ECOLOGIST'S AND ECONOMIST'S VISIONS OF THE WORLD

The ecologist's and the economist's visions of the world can be reconciled, in the case of soils, by recognizing that the soil resource is irreplaceable in both worlds and that it plays an essential role in both pyramids. Whether your preferred vision of the world has the pyramid right-side-up or up-side-down, conserving the soil resource will improve the stability and yield of both pyramids. Technological advances (in both conventional and organic farming) are welcome and needed, but they will not replace soil.

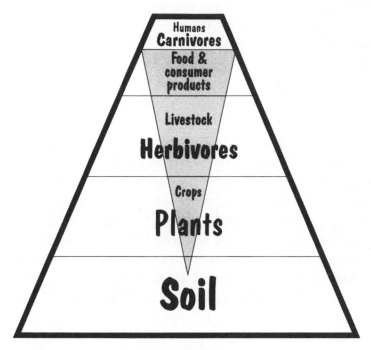

A Common View of How the World Works:
Economist's Pyramid Within Ecologist's Pyramid

The human economy works within the confines of the natural economy. The ecologist's and economist's pyramids can remain in their present orientation as long as the economist's pyramid is placed within the ecologist's pyramid. The human economic system can function pretty much as modern neoclassical economic theory describes it, except that the economic system cannot stand by itself. It is part of a larger system, which is represented by the ecologist's pyramid. Although largely unpriced and thus absent in economic analyses, natural resources, such as soil, water,

> It of course goes without saying that economic feasibility limits the tether of what can or cannot be done for land. It always has and it always will. The fallacy the economic determinists have tied around our collective neck, and which we now need to case off, is the belief that economics determines all land-use. This is simply not true. An innumerable host of actions and attitudes, comprising perhaps the bulk of all land relations, is determined by the land-users' tastes and predilections, rather than by his purse. The bulk of all land relations hinges on investments of time, forethought, skill, and faith rather than on investments of cash. As a land-user thinketh, so is he.
> —**Aldo Leopold**

air, forests, and oceans, provide essential functions and hence must be viewed as an integral part of our economic system.

Many farmers whose families have lived on the farm for several generations have a land ethic—a reverence for the land that supports them. This ethic, and the desire to hand over their farm to their children and grandchildren, can provide the incentive to manage the land wisely and to prevent soil degradation. Even these farmers, however, must balance their land ethic with their families' pressing financial needs. In many places, corporate farms are replacing the family farm, and it is less clear that the long-term view of a land ethic is compatible with the bottom line of a corporation's quarterly or annual profit sheet. In many regions of the world, farmland is being expanded into previously unsettled regions by people migrating from overcrowded areas, and these landless poor have not yet developed a land ethic. Although the land ethic supports sound management, it is probably not enough to protect our soils for productive agriculture in both the near and distant future. Farmers also need economic incentives to practice erosion control and prudent irrigation.

In the United States, there have been some successes in encouraging farmers to take easily erodible land out of production through a government incentive program called the Conservation Reserve Program. If the farmer agrees to take the most vulnerable land out of agricultural production and plant it in native grasses or trees, then he can participate in various price support and credit programs. Despite considerable success in reducing rates of soil erosion, these programs were nearly eliminated by Congress in 1995 because of opposition to environmentalism in general, and specifically to the idea of government telling farmers how to manage their land.

In the following chapters, examples illustrate how new initiatives that draw upon the ideas of ecological economics are finding ways to include the ecological values of resources like soils in the economic marketplace system. In the United States, the Conservation Reserve Program is expanding again. Another approach, conservation easements, allows landholders to take tax write-offs in exchange for easements that stipulate the type of development and management allowed on the land in perpetuity. New ways of planting with only minimal tillage or no tillage, which helps to minimize erosion, have been encouraged by government-supported research and are becoming economically attractive to farmers. Tax credits for planting windbreaks, building terraces, installing drip irrigation, and other investments of capital and labor are needed to further reduce the rate of soil erosion and to

avoid salinization in many regions and countries throughout the world.

These methods of encouraging sound soil management
are still evolving and many are controversial, but we no
longer live in a world where a farmer should boast of wearing out three farms in his lifetime, even if such actions
were economically rational in the short term. There are
too many mouths to feed and too little land to cultivate to
permit such recklessness, although it is occurring nevertheless. The disciplines of ecology, agronomy, and economics must be engaged and integrated to find ways to
give farmers of the twenty-first century the incentives and
technology needed to act in a manner that is both economically and ecologically rational.

3 🌿

The Price Is Wrong

Advantages and Dangers of Cost-Benefit Analysis

Long an adherent to the "drive-it-'til-it-drops" school of car ownership, my frugal Scottish upbringing was put to test when I read an impressive analysis of the safety features now available on new cars. Unlike my old 1986 model that I bought just after finishing graduate school and that was still running fine, the new cars have all-wheel drive, antilock brakes, air bags, and side impact panels. Now that I am a father and no longer live on a graduate student's stipend, and recognizing (as my Scottish grandfather used to say, but only half believed) that "you can't take it with you," I considered parting with some of my money to obtain these safety benefits. To decide whether this was a smart thing to do, I conducted my own cost-benefit analysis. Hanging on to my old reliable car a few more years was clearly the option with the lowest cost. The

37

tricky part was figuring out how much the benefits of a new car's safety features would be worth.

Consumer Reports magazine provided some insight. For the prospective buyer who was considering which options to include in a new car package, they argued that antilock brakes and air bags have demonstrated effectiveness and are worth the extra money. On the other hand, "adjustable ride control," whatever that is, was reported to be "more gimmick than benefit."

Thanks to the *Consumer Reports* cost-benefit analysis, I knew which safety features were worth getting, but that did not answer the question of whether I should buy a new car. I knew that if I really wanted to do the cost-benefit thing right, I would have to find statistics on the probability of getting in an accident per mile driven and the probability that these safety features would save life or limb during such accidents. I would then have to calculate how many miles my family drives and how many dollars we consider our lives and limbs to be worth. Looking at my six-year-old son, that value seems incalculable, although I must admit that it also seems to vary according to his mood of the minute. Then it occurred to me that I could also buy a car with a CD player, and I wondered if I should include in my analysis the benefit of the stress reduction I would enjoy from boogying to some bebop in my new buggy.

In the end, I gave up on this daunting analysis. I bought the car, the way everyone else does, based on my parental instincts that it was the right thing to do. Perhaps most of us should not expect to complete strict cost-benefit analyses for our personal purchases and decisionmaking, but government decisions that affect millions of people and that consume billions of tax dollars surely should be based on the most sophisticated and objective cost-benefit analysis available, right? After all, economists are trained to calculate these complex probabilities and the value of things. Well, maybe. Should we trust others, especially those with predominantly neoclassical economic training, to calculate the value of the universe of things that affect our lives and our children's lives?

We have already seen that neoclassical economics does not give much value to dirt cheap soil, and the following chapters demonstrate that groundwater and air are also usually unpriced or underpriced. If the cost-benefit analysis is applied to a decision about conserving one of these essential resources, will the full benefits of their conservation be included in the calculus of the sophisticated econometric models? In most cases, many of the benefits of environmental protection are left out of the balance sheet entirely because they are too difficult or impossible to calculate by standard neoclassical economics. In contrast, the costs of preventing soil erosion, of wastewater cleanup, and of taxing oil to promote energy conservation are relatively easily calculated, and they make their way into the cost column of the analysis. With the costs easy to tally and the benefits often practically incalculable, is it any wonder that the environment usually comes out on the short end of cost-benefit analyses?

I have been referring to costs of implementing environmental protection and the benefits of doing so. Conversely, some projects entail economic benefits and environmental costs. In the case of constructing a dam, for example, the benefits that can be reasonably well calculated by economists may include flood control, irrigation water, hydroelectric power, and recreational boating. The readily calculated costs include construction costs and interest on construction bonds. The less easily calculated costs include environmental destruction, such as loss of river habitat, downstream erosion, and the costs to future generations of decommissioning the dam and reclaiming the landscape when the reservoir eventually becomes filled with silt. So sometimes the environmental perspective is in the benefits column (benefits of protection) and sometimes it is in the cost column (costs of destruction), but in either case, the monetary values are usually very difficult to evaluate.

COMMON-NONSENSE COST-BENEFIT ANALYSIS

Recent efforts by anti-environmentalists in the U.S. Congress to require the Environmental Protection Agency to base all of its environmental regulations on cost-benefit analysis may sound reasonable at first, but beneath this sheep's clothing you can bet that there is a wolf who will make short-term financial gain from environmental destruction. The history of cost-benefit analysis of the environment is checkered, at best, and demonstrates how young this "science" is.

There have been some strange entries in the environmental benefits column of cost-benefit analyses. In the early 1980s, policymakers in the United States were debating the wisdom of reducing air pollution that causes acid rain (and the debate continues today). The most well-established effect of acid rain is acidification of mountain lakes and the resulting loss of fish populations. In one analysis, the benefits of reducing acid rain were calculated from the average number and weight of fish that sports-

What is a wild goose worth? I have a ticket to the symphony. It was not cheap. The dollars were well spent, but I would forgo the experience for the sight of the big gander that sailed honking into my decoys at daybreak this morning. It was bitter cold and I was all thumbs, so I blithely missed him. But miss or no miss, I saw him, I heard the wind whistle through his set wings as he came honking out of the gray west, and I felt him so that even now I tingle at the recollection. I doubt not that this very gander has given ten other men a symphony ticket's worth of thrills.

—Aldo Leopold

men caught in the Adirondack mountain lakes of New York, multiplied by the cost per pound of fish in the local supermarket. Nice try, but this economist was obviously not a fisherman, or he would have valued the one that got away more than the ones in the creel (and don't forget that the one that got away is always bigger). There is more to the value of life in a lake than the fish that are caught, and there is more value to fishing than the reduced grocery store bill.

If a fish is hard to value, try valuing a human life. Global warming is expected to cause increases in human diseases, because the habitat for mosquitoes and snails and other vectors of tropical diseases will expand. Higher temperatures in urban areas will also increase the rate of photochemical reactions in the air that create smog, causing increased respiratory illness. Although some of these diseases, like malaria, are already spreading to the United States, most of the loss of life will occur in poor countries that have fewer resources to cope with the increased incidence of disease. Malaria and schistosomiasis are most common in these tropical countries, and the worst air pollution problems are also currently in large cities of developing countries, like Mexico City and São Paulo.

In one study that attempted to calculate the benefits of avoiding this increase in disease due to global warming, the number of lives saved were estimated and a dollar value per life was based on a person's earning potential. Because Americans, Canadians, Europeans, and Japanese earn more money than do Brazilians, Mexicans, Chinese, and everyone else, the value of an American life was placed at $1,500,000, whereas the value of a Mexican life and the lives of other citizens of third-world countries was estimated at $100,000. Not too surprisingly, people from Mexico and other developing countries took offense at being devalued by a factor of fifteen. Because much of the human suffering due to global warming will occur in developing countries, the estimate of the benefit of avoiding those deaths depends greatly on whether those lives are given a "third-world" value or a "first-world" value. If every life were valued at $1.5 million, then a cost-benefit analysis would be more likely to support spending money now to avoid global warming. But if most of the people who would be dying were worth only one-fifteenth as much, then the cost-benefit analysis might tell us that we would be better off saving our money. Is earning power, average amount of life insurance, or some other economic indicator an appropriate measure of the value of a life, and does the value depend on the person's nationality, place of residence, or social status?

Some economists respond that this valuation of human lives has been misinterpreted. They try to use objective means of calculating the value of a "statistical" life in a country or society by estimating people's willingness to pay to avoid the probability of a "statistical" death. Because Americans have more disposable income than many other people of the world, they tend to be more willing to pay for safety features such as auto air bags, which are relatively expensive per death avoided. Hence, the calculated value of a "statistical" American life is high. Most Haitians could not afford and would not be willing to pay for air bags. If they had the money, they could invest it in more cost-effective death avoidance. A lot of deaths could be avoided in a place like Haiti for a relatively small investment in food, basic health care, and sanitation, and so the cost of avoiding a statistical death is low there. Nevertheless, human suffering is equally painful whether it occurs in Haiti or the United States and regardless of how much each society is able and willing to pay to avoid it. Hence, cost-benefit analyses that equate human death and suffering to a monetary value, using economists' theoretically correct calculations of the value of "statistical" lives and death avoidance, remain controversial and repugnant to many.

Economics does not do well in determining the value of the lives of either fish or humans. Likewise, the value of the whole environment, including the direct benefits that humans obtain from the environment, is very difficult to measure, but a new breed of economists is now trying to do exactly that.

ECOSYSTEM SERVICES

Ecologists use the term *ecosystem* for a grouping of plants, animals, and microbes that interact in systematic ways with

their environment. Most ecosystems, like forests, grasslands, and wetlands, have effects on their own environments, such as purifying the water and air that pass
through them. They provide habitat for the insects that
pollinate flowers of both crop plants and native plants.
They absorb rainfall, which reduces flooding and erosion.
Their soils provide nutrients and water for plants. The
plants, animals, and microorganisms that make up an
ecosystem are the original sources of many valuable pharmaceuticals, including most antibiotics used today. These
beneficial effects of ecosystems on human well-being have
recently been dubbed *ecosystem services.*

Ecological economists are trying to determine monetary
values for the ecosystem services provided by soils, wetlands, native forests, oceans, and all types of ecosystems.
Traditional marketplace pricing does not work well for
these natural resources because they are not currently
traded efficiently in the marketplace. Instead, the ecological economists try to identify the services provided by soil,
assess a monetary value for those services, and then calculate of the dollar value of the service provided per ton or
per acre of soil. Likewise, forests and wetlands can be
viewed as systems (ecosystems) that filter and purify the
water flowing through them, thus providing a service that
can be roughly equated with the cost of building and running a municipal water purification plant.

New York City recently chose to invest $1.5 billion for
protecting and restoring forested watersheds in the
Catskill Mountains, which is the source of much of the
city's municipal water supply, rather than paying $6 billion
to build, and $300 million annually to operate, a new water
filtration plant. If the forested watersheds are maintained

in good health, then they will provide good quality water, and the filtration plant will not be needed. In this case, the monetary value of the clean water provided by the Catskill watersheds can be calculated reasonably well, and it turns out to be quite a bargain.

Although New York City could calculate that investing in its upstream forests is cheaper than investing in a new water purification plant, calculating the value of many other natural resources, such as soil, is more difficult and controversial. A farm with very fertile soil may be worth more than a farm with infertile soil, so market forces are not entirely absent. But soils are left out of accounting sheets that show entries such as the costs of seeds, labor, and fuel for machinery, the income from sales of the harvest, and depreciation of the tractors and farm buildings. Costs may be kept low and short-term profits high by ignoring the need for soil conservation expenditures, such as planting rows of trees that act as windbreaks. The value of the soil lost to wind or water erosion could be thought of as a type of depreciation, similar to depreciation schedules allowed for equipment and buildings, but soil is not included in standard accounting depreciation schedules, and perhaps it should not be treated this way. A worn out, fully depreciated tractor or any other standard capital investment can be replaced at any time if there is enough money. The soil, in contrast, cannot be so simply replaced when it is worn out. Rather, soil is a type of *natural capital* that is virtually irreplaceable and, therefore, invaluable. At the same time that we try to "get the prices right" for natural resources and ecosystem services, we must keep in mind that when the value of a resource is unmeasurable by economists' tools, it is not *un*valuable, but rather it is *in*valuable.

After studying numerous types of ecosystems around the globe and examining how they affect their surrounding environs in ways that provide beneficial services to humans, one group of ecological economists recently calculated that the worldwide value of these natural ecosystems services is about $33 trillion. This enormous sum is bigger than the mid-1990s global GNP of about $18 trillion, which is the sum of all of the marketplace goods and services calculated by neoclassical economics. Moreover, because of the nature of the uncertainties in their calculations, these ecological economists chose a conservative appraisal, so that they probably underestimated the true value of all natural ecosystem services.

Although we must admit that the best estimate of the monetary value of ecosystems and the services they provide humanity is rather uncertain, we are certain that the true value is big—many trillions big, at least as big as the global GNP, and probably much bigger.

Ecological economists have placed the value of ecological systems into perspective by making a commendable first cut at monetarily assessing the essential ecosystem services that have not been included in neoclassical economics. There are obstacles, however, to utilizing these preliminary estimates. First, the large dollar values (e.g., $33 trillion) calculated for ecosystem services by ecological economists must withstand the scrutiny of the evolving field of economics. These estimates have certainly stimulated a lot of debate among economists, but a consensus has not yet been reached.

Second, these new values of ecosystem services must be accepted and effectively incorporated into cost-benefit analyses when decisions are made regarding draining of wetlands, building or dismantling of dams, managing forests and watersheds, and investing in soil conservation practices. Then, perhaps, more of these ecosystems and their services might be maintained for their economic value to humanity. Instead of a wetland being seen as a worthless swamp, for example, it is seen as providing considerable economic benefit by purifying the water that passes through it, which then supports economically important fish spawning grounds in downstream estuaries and provides potable water for downstream communities. The monetary benefits of ecosystem protection can then be shown on economic balance sheets to be greater than the costs.

COST-BENEFIT ANALYSIS FOR INFORMED DECISIONMAKING

The recent anti-environmental sentiment in the U.S. Congress stems from a perception that some governmental regulations have been excessively restrictive and costly. One could certainly imagine that an overzealous regulator in green underwear might not consider the economic costs of a proposed regulation unless the regulatory process required something like a cost-benefit analysis. So cost-benefit analysis in some form is probably justified, but how can we make sure that the long-term benefits provided by natural ecosystems, many of which are *in*valuable, get their due credit in these analyses?

I believe that the answer is not very different from the way I bought a car. First, as much data should be gathered as possible. I stopped after the *Consumer Reports* article, but professional analysts would be expected to dig deeper.

The cost-benefit analysis should be conducted using the best tools of neoclassical economics and ecological economics, but the results should be weighed and interpreted by good human judgment.

Although some of the entries in the columns for benefits of environmental protection and costs of environmental destruction may be unconventional, the cost-benefit analysis can be constructive as long as the inadequacies of the monetary estimates of ecosystem services are recognized. Considering the estimated monetary values of the environmental costs and benefits in addition to conventional economic costs and benefits can be enlightening, albeit incomplete and less than exact.

Legislation requiring that the decision to regulate pollution of the environment be based on a strict formula of whether economic benefits exceed the monetary value of costs of environmental damage would be unwise. The costs and benefits of environmental protection are often unmeasurable or are very imperfectly measured by economic standards, and so strict formulas based on such inadequate data are equally inadequate.

Likewise, we must be on the alert for efforts to influence decisionmaking that are hidden under the mantle of apparent objectivity of cost-benefit analyses. With so much uncertainty in estimates of the monetary value of environmental benefits, individuals, corporations, government entities, and environmental advocacy groups can "stack the deck" in a cost-benefit analysis by changing a subtle assumption here or there so as to obtain their desired outcome. Underlying assumptions in cost-benefit analyses may reflect the subjective values of those conducting the analysis, and so the assumptions must be sifted out and stated openly and clearly. Another risk is that the findings of cost-benefit analyses conducted under the guise of sophisticated econometric analyses are often difficult for the rest of us to understand and thus challenge.

In the end, the decisionmakers, be they government regulators issuing a regulation, members of Congress voting for or against a bill, or consumers deciding to buy a gas-guzzling utility vehicle or fuel-efficient sedan, will base their decisions on a personal judgment of what is right. The cost-benefit analysis provides some data to help reach that decision, but it is imperfect, and a human being is ultimately responsible for making the decision. Our values, including those that we can express accurately in dollars and cents, and those that we can express only in common sense, guide our decisions.

CHOOSING YOUR OWN POISON AND OTHER TRADE-OFFS

When an army doctor at a battlefield has a limited amount of medicine and bandages, the concept of triage requires that the medicine be used for those who need it most, that those who will survive without it can do without, and that it should not be wasted on those who will die anyway. Due to lack of money, triage is being applied to the environments of U.S. military bases across the country, where many of the worst toxic waste dumps of jet fuel, solvents, and nuclear and chemical contaminants of various kinds are found. Given limited taxpayer willingness to pay for the bandages and medicines needed to clean up these toxic waste dumps, some sort of priorities have to be developed to put the buck where it will get the best cleanup bang. A cost-benefit analysis might help determine which military base mess to aim at first.

The notion that some messes must be tolerated, at least temporarily, because of limited resources to clean them up gives many environmentalists the heebie-jeebies. Similarly, some purists accept no "unnatural" substances in food. Organically grown produce is *de rigueur* for many people. The question arises: Is there room for a little bit of poison in the environment—be it the less urgent toxic dumps or the barely detectable residues of pesticides on apples in the grocery store? So-called natural products, like peanuts that are contaminated by mold during storage and comfrey tea, are carcinogenic if consumed at high enough levels. Then there are coffee, alcohol, tobacco, marijuana, and other mind-altering compounds that are produced from natural products. If nature is full of substances of dubious qualities for human consumption, then why not permit man-made compounds that might be relatively safe when used in moderation?

Cost-benefit analysis may help define these issues, although it cannot resolve them. The analysis is doubly tricky because some of the costs of environmental protection are not only the traditional negative effects on GNP, but also include concerns of maintaining an abundant supply of nutritious food at an affordable price. Food pricing is extremely complicated, with most governments throughout the world subsidizing various crops in numerous direct and indirect ways for all kinds of economic, social, and political reasons. Ignoring much of that complexity for the moment, one can argue that the proven benefits of fruits and vegetables in the diet argue strongly for keeping their price low so that people at all income levels can afford to buy them and eat them. A bit of comparison shopping in your local supermarket will quickly show that organically grown produce often (but not always) costs two to three times as much as conventionally grown produce. If we immediately prohibited all pesticides in agriculture, the price of produce might rise to the point where people would eat fewer fruits and vegetables and thereby suffer the deleterious consequences of an improper diet.

When my son was about two years old, he drank the equivalent of his body weight in apple juice every month, which we preferred to have him drink rather than soda. I was concerned, however, that a small amount of a harmful pesticide residue in his apple juice would add up to a significant dose at the rate he drank the stuff. What's a father to do? Clearly, there is a role for comparing the risk of consuming potentially harmful pesticide residues on food that is otherwise cheap and nutritious against the risk of avoiding nutritious foods in healthy quantities because they seem too expensive when grown completely organically.

In the late 1980s, a scare developed over the use of Alar, a chemical that apple growers sprayed on their orchards to make the apples turn red and ripen simultaneously, which improved the ease of harvesting as well as the visual appeal of the apples. The Environmental Protection Agency had conducted studies to determine the health risks associated with eating apples treated with Alar, but the results were equivocal. The Natural Resources Defense Council (NRDC), a national environmental advocacy group, differed with apple growers on the interpretation of the data from these studies. Some have argued that the NRDC overestimated the risk by several orders of magnitude; however, the NRDC has defended its analyses, especially for the exposure levels of children, who consume large quantities of apple juice relative to their body weights. We will probably never know who was right, because Alar has since been taken off the market by its manufacturer, and no one is conducting the studies that would resolve the debate. For adults who ate an occasional apple, the health risk associated with the ingestion of Alar residues was probably not very high, but for children who drank a lot of apple juice, like my little boy did (and still does), a significant health risk was plausible.

Although most of the debate was about the details of the risk analyses—the amount of Alar that could be "safely" consumed per pound of child—the story that struck me was that Alar was used primarily for cosmetic purposes so that nice red apples could be marketed. My son's favorite video used to be *Snow White,* and he already knows not to judge the goodness of an apple by its redness. Alar was not in the same league of danger as the evil stepmother queen's magic potion, but exposing people to even a small risk for the sake of cosmetic appeal of the fruit does not make the industry look like the fairest in the land. Unfortunately for both growers and consumers, the scare caused a sudden temporary decline in apple sales. The scare has long since passed, Alar has been removed from the market, and both apple growers and apple eaters seem to be living reasonably happily ever after without Alar, without much, if any, change in the affordability of apples. Well nourished on substantial quantities of apple juice and other nutritious foods, my son is growing up like a healthy and handsome prince.

Alar is just one of hundreds of pesticides in use in agriculture throughout the world. Americans rely on the Environmental Protection Agency, the Department of Agriculture, and the Food and Drug Administration to regulate the use of pesticides. Unfortunately, the budgets for inspections are inadequate, resulting in only a tiny fraction of the produce being inspected. Because they are often exposed to large doses while applying the pesticides in the fields, agricultural workers suffer the greatest risks.

Many of the pesticides that the United States has banned for use at home are still being manufactured in the United States and exported to other countries, where their use is largely unregulated. Ironically, some of these pesticides then make their way back into the United States as residues on imported produce. The government inspects only a small fraction of the produce as it is imported, and thus our control over pesticides in the food supply is currently questionable.

An agricultural practice called *integrated pest management* (IPM) uses knowledge of the life cycles of pests to control their populations with appropriate crop rotations, timing of planting, natural predators, and when necessary, judiciously applied minimal doses of pesticides. Although the effectiveness of IPM has been well demonstrated on experimental farms, its application by commercial growers has been disappointing, probably because they base their decision to use or not use pesticides primarily on the bottom line of their accounting sheets, which does not include the nonmonetary costs of pollution from pesticides. More widespread adoption of IPM practices will require either economic incentives for the farmer, such as taxes on pesticides, or greater regulation. For government to intervene with either regulations or economic incentives, some sort of cost-benefit analysis would probably be needed. If the benefits of reduced pesticide exposure to farmworkers, consumers, and wildlife are properly included in these cost-benefit analyses, then IPM may finally get the attention that it deserves.

Where should the line be drawn between the "danger-ous" and the "acceptable" use of pesticides or any other chemical in the environment? The answer to this not-so-rhetorical question is a matter of judgment. Data on costs and benefits help to put this judgment call into perspec-tive, but these decisions cannot be determined by rigid formulas. I do not use herbicides on my lawn, and it does not look as pretty as my neighbor's lawn, but I rest easy knowing that my dog and my child can play there without risk of exposure to potentially harmful pesticides. That peace of mind is more important to me than a beautifully manicured lawn, but my neighbor apparently does not share my opinion. Much to the dismay of some of my envi-ronmentalist friends, however, I have used herbicides on poison ivy growing on the periphery of my property, be-cause I prefer the risk of that man-made chemical over the risk of my son's exposure to the natural chemical in poison ivy before he is old enough to identify "the leaves of three to be let be." Those are my values. Likewise, when govern-ments make decisions, they must reflect the values of their citizens and not hide behind the guise of seemingly objec-tive and sophisticated, but usually inadequate, cost-benefit analyses.

MORE THAN A GAME OF *The Price Is Right:* THE MARGINAL VALUE OF MARGINAL ANALYSIS

As discussed in Chapter 1, the market economy usually does a good job of pricing the goods and services that are exchanged within the marketplace—be they handbaskets, hula hoops, magazines, or mangos—because these goods and services are efficiently traded within the economic system. Perhaps ecological economics will succeed in assessing appropriate monetary values to the essential goods and services that nature provides us but that are not currently traded efficiently in our marketplace and are not part of the GNP. Protecting our essential environmental resources, however, is more than a game about getting the prices right. In addition to assigning dollar values to soil, fresh water, air, forests, and oceans, we must recognize that our most basic natural resources, no matter how accurately or inaccurately they are priced and traded, are essential, irreplaceable, and nonsubstitutable.

Here lies another crucial difference between the economist's and ecologist's vision of how to value resources. Economists focus almost exclusively at what they call the *margin*. Economists seldom try to value the entire stock of a resource, such as all the trees in the world, all of the water underground, or all of the fish in the oceans. Instead, they focus on how, for example, the price of lumber might decrease if logging activity in one region were to increase the supply of sawlogs. Prices are determined in an efficient market by these changes in supply and demand and often apply to only a small part of the entire resource. Only a small fraction of the world's trees is being bought and sold on the market at any point in time, and it is this small fraction, at the margin of the larger forest resource, for which the marketplace provides prices. The ecological economists' estimate of $33 trillion as the value of global ecosystem services also used marginal values of individual ecosystem services and summed them up across the globe. The validity of this approach has been questioned by neoclassical economists, and it continues to be debated. Cal-

culations of GNP are based on marginal values, and GNP
assesses the value of those goods and services that are actually traded *at the margin.*

In the hypothetical case that deforestation proceeded to
the point where trees became very scarce, the remaining
trees would increase in monetary value as the supply of
wood products had trouble keeping up with the demand.
The last few standing trees would have extremely high value
in the economist's world, either for their very rare wood or
as tourist curiosities. The ecologist, in contrast, is keenly
concerned about the value of the entire forest resource of a
region or of the globe. How do forests affect the quality and
quantity of a region's water supply? How do the forests of
the world regulate the global climate and support biological
diversity? Although the last few trees in an almost completely deforested hypothetical world would have very high
monetary value for the economist, they would have almost
no ecological value. A small stand of forest can no longer
provide sufficient habitat, water purification, and regulation
of climate to make a difference in these essential ecosystem
services. Ironically, we must have big tracts of forest to provide habitat, clean water, and a habitable climate, but the
more abundant the trees are, the less value they have as timber "at the margin" in economic analyses. Assigning accurate prices to the nontimber values of forests would help address this conundrum (i.e., getting the prices right for the
marginal values of habitat, clean water, and climate moderation), but pricing these nontimber products and services at
the margin may not be enough. The value to humanity of
the many products and services provided by large expanses
of forests may be greater than the sum of the many small
parts calculated at the margin.

Cost-benefit analyses are about trade-offs, and they depend upon values at the margin. For example, how many acres of forest can we do without in a local area in order to increase some other form of economic output, such as a strip mall or a strip mine? These are changes at the margin of the forest resource. The analysis assumes that a marginal loss in forest cover can be traded for and substituted with a marginal increase in monetary capital or human capital. It also mostly ignores the cumulative effects of many such marginal trade-offs as they add up to large-scale deforestation. Marginal changes are well suited to traditional economic analyses such as short-term local-scale cost-benefit analyses. Conversely, these economic analyses are not currently suited to addressing the issue of how much of a natural resource must be maintained at regional or global scales.

In Chapter 8, in discussing the protection of habitat for the diversity of plants and animals, I suggest that half of the land that was once forested—before the recent wave of population growth and deforestation—should always be maintained and managed as forests. That conclusion cannot be reached by analyzing the value of forests at the margin. Rather, I argue that big chunks of forest are needed to maintain the climate, groundwater resources, and habitat for a diversity of plants and animals. My rough estimate of 50 percent may be wrong—perhaps we need 70 percent of the once forested area to remain in forest, or perhaps we could get by with 40 percent. Nevertheless, my point is that diddling around at the margins will not establish the value of forests that must be recognized on regional and global scales. Clearing forests, a few acres at a time, can be rationalized as promoting economic productivity, but it has

> The economy is a wholly owned subsidiary of the environment.
>
> **—Gaylord Nelson, former U.S. senator of Wisconsin**

a way of adding up to large-scale deforestation with regional and global consequences that were not a part of the local-scale cost-benefit analysis. The right prices for ecosystem services, and the marginal valuation method upon which these are based, are useful tools for many inevitable decisions about trade-offs, but they may be inadequate to conserve the natural resources that we will need at large scales for long-term economic and ecological prosperity.

Will cost-benefit analyses continue to provide economic justification for whittling away at the margin of the forest until the forest is too small to provide the essential ecological and economic services of providing clean water, wildlife habitat, and climate moderation? This seems to be the direction we are headed (current rates of deforestation are discussed in Chapter 8). Until economists can adequately address this question, we have yet another reason to regard cost-benefit analysis with healthy suspicion.

In the chapters that follow, I examine in greater detail how we cannot do without forests, wetlands, uncontaminated groundwater, or a habitable climate. These resources are the underlying basis of our wealth and prosperity. Cost-benefit analysis, when used appropriately, may be a useful tool to help put into perspective the value of ecosystems as they are protected, contaminated, or destroyed, and as trade-offs are rationalized at the margin.

But equating ecological systems with economic systems strictly on a dollar-for-dollar basis misses the point that the economic system cannot exist without the ecological system. The economic pyramid is always contained within the larger ecological pyramid. The ecological system must remain healthy at all scales—local, regional, and global—if the economic system is to survive.

4 🌿

Future Shock Discounted

Another Devil in the
Details of Cost-Benefit Analysis

WHERE I WENT TO SCHOOL in North Carolina, the farms are often intermixed with forests, and most farmers have a piece of land that they call the "back forty." The farmers have left these forty acres or so in forest. They and their families use these woods for hunting, as a source of firewood, as a hedge against the need for more cropland later, or simply as a peaceful place to take a walk. One of the forestry extension agents at the university spent most of his time counseling farmers on how they could manage these small forest stands for future timber production. His sales pitch was that the back forty, when properly managed for growing economically valuable trees, could act like an individual retirement account (IRA) that would provide needed income and security when the farmer is ready to retire. Managing the forest for the growth of merchantable timber is only part of the process, however; knowing when to sell is also critical. Ignoring the hunting, firewood, and peaceful walks for the moment and assuming that retirement income is the main objective of managing the back forty, it might make more sense to cut the trees, sell the logs, and invest the proceeds now. Money in a real IRA invested in mutual funds might grow in value faster than the trees. Calculating when to sell timber is one of the first things that forestry students learn in their introductory forest economics courses (again, ignoring the nonmonetary value of the back forty). The key to the purely economic analysis is the application of a *discount rate,* which allows comparisons of the value of timber sold now to timber sold in the future.

I had studied natural resources for twenty years before I realized that the same discount rates that I had learned to apply to timber sales are also being applied in virtually every economic cost-benefit analysis that involves management of environmental resources. As a result of this discounting, the policy decisions based on economic analyses have been putting less value on the future than on the present, and most of us have been unaware that this was happening. According to neoclassical economics, the old adage that "an ounce of prevention is worth a pound of cure" may or may not be true, depending on what discount rate is used. It is extremely important that we understand discount rates and how they are used and misused to devalue the future environment.

A Primer on Discounting

Fortunately, discounting is not terribly complicated, and a simple lesson of what discounting is all about can be learned quickly. Let's assume that the timber on your parcel of land is worth $1,000, after paying the logging costs;

you could cut down the trees and invest the $1,000 pro-
ceeds. If the current annual rate of return on that investment is 8 percent, then that investment will be worth
$1,080 a year from now. On the other hand, you could wait
a year before cutting the forest, in hope that it will grow
and be worth more. For this example, let's say that the
profit from cutting the slightly larger timber will be $1,050
when cut and sold a year from now. We can already see, by
comparing $1,080 and $1,050, that you are better off cutting now and investing the profits.

Economists use a somewhat more complicated method,
but one that yields essentially the same answer. They apply
a discount rate to calculate the dollar yield of each of these
two options in today's dollars, rather than next year's dollars as I have done above. The value of timber delivered at
a future time is discounted to account for the lost investment revenue that would have been earned had the timber
been cut sooner and the profit invested. In this case, cutting now yields $1,000 in today's dollars. Cutting a year
from now will yield $1,050 minus a discount rate that is
usually chosen to be the current common annual rate of
return of monetary investments, which in this example is 8
percent. So the timber delivered a year from now by delaying the cut is discounted by $84 ($1,050 x 0.08), making it
worth $969 ($1,050 − $84) in today's dollars. You can see
that the answer is the same—the timber is worth more in
today's dollars if you cut it now ($1,000 versus $969), assuming an 8 percent discount rate.

In some cases, where the trees are growing past the
threshold that separates less valuable (small) pulpwood
used for paper making to more valuable (larger) sawlogs
used for furniture making, the value of the timber in-

creases more than the discount rate and the wait is worthwhile. In most cases, however, economic analyses favor
cutting now rather than later, because rates of growth of
relatively mature trees seldom match the rate of return of
monetary investments.

Let's take a second example, using heating oil. A heating
oil company offers you a deal in which you can buy 100
gallons of heating oil delivered this year for $100, and if
you pay in advance, they promise to deliver another 100
gallons again next year if you pay only an additional $95
now. So for $195 paid now you get 200 gallons of heating
oil, half of it delivered this year and half next year. Is that a
good deal? Why are they willing to sell next year's oil at a
cheaper price? For simplicity, let's assume that the cost of
oil is stable—that is, that we can be sure there will not be
any oil embargoes or other geopolitical events that will
cause the price to vary (of course, this assumption is not
necessarily true, but the assumption greatly simplifies the
explanation of discounting). In effect, then, the oil company is discounting the price of oil delivered a year from
now by 5 percent (i.e., $95 is 5 percent less than $100).

Your alternative to taking them up on their offer is to
buy only the 100 gallons that you need this year and to invest the $95 that you would have paid for next year's oil. If
you invest that $95 in a high-yield mutual fund that does
well, it might be worth $105 by next year. In that case, you
could buy 100 gallons of oil next year at the usual undis-
counted price of $100, and you would still have $5 left over
to buy some wool socks. On the other hand, if you are a
conservative investor who does not like to take risks, you
might put your $95 in your savings account at the local
bank, where it will earn only about 2 percent interest.

Earning a little less than $2 in interest will make your investment worth only about $97 after a year. In order to buy $100 worth of oil next year, you'll have to cough up another $3 from somewhere else. You would have been better off taking the oil company up on their deal to pay for the second year's oil in advance for only $95.

The oil company can make its offer of a discount on next year's oil because they calculate that they can invest the $95 dollar advance payment from you and that they will earn more than 5 percent on their investment. Perhaps they need capital to expand their profitable business, perhaps they can take advantage of economies of scale that provide better returns than those available to individual investors, or maybe they are just especially savvy investors. Whatever the reason, they are confident that they will make more than a $5 return on investing your $95 advance payment, so they can afford to deliver $100 worth of oil to you next year. This sounds like a good deal for everyone involved—the oil company has obtained the capital it needs to make new investments that will improve their profits, while you get a discount on your oil that is a better return than what you could get in your bank's savings account.

WINNERS AND LOSERS

In these two examples, the monetary winners are the farmer who cuts his forest now, the homeowner who buys discounted heating oil, and the oil company. The losers in first example are the environment and, depending upon their values, the farmer's children and grandchildren. Cutting the back forty woods now may make the most sense for maximizing the value of a high-yielding monetary investment that the farmer may use in retirement or that may be inherited by his descendants. But the farmer and his children and grandchildren will have to wait for many years before a regrowing forest will again supply firewood, game, recreation, and aesthetic solitude. The environment and the future farmers also lose because the role of the back forty woods as a windbreak to prevent wind erosion of soil on the farm, as a buffer strip to prevent pollution of streams, and as habitat for animals will also be lost for several years or decades. Although the children and grandchildren may inherit more money from the monetary investments, they may also inherit a farm with less fertile soil, polluted streams, and fewer opportunities for recreation and peaceful enjoyment. More importantly, the children and grandchildren have no voice in that decision but must accept the consequences of their parents' decision today.

A similar real-life example comes from the forest where I do research in Brazil in the eastern Amazon Basin. We pay rent to a rancher so that he will leave a patch of intact forest on his ranch for us to study. We make measurements of the soils and vegetation in cattle pastures and young and old forests on this ranch. As a result, this patch of a few hundred acres of mature forest is one of the few left standing in this logging and ranching region of the Amazon Basin. We once tried to convince the rancher that he should preserve that forest in perpetuity, so that his children and the school children of the community would be able to visit and study a nearby rain forest. Ironically, most of the children in this part of the Amazon Basin have never been in a forest. We offered to develop an educational program for the schools to make good use of the educational value of this forest. We hit a brick wall. The rancher said that he is more concerned with leaving money to his children than leaving a patch of forest. This savvy rancher has also justified raising our rent by using a high discount rate when calculating his lost revenue from delaying harvest of the forest.

When our research project is over and we stop paying rent, that forest will probably disappear, with the proceeds from the logging potentially increasing the monetary inheritance of the rancher's children. I do not know if the children will also inherit their father's preference for money, but they have no choice in the matter. The school children of the community will certainly lose an opportunity to learn about the ecosystems that once inhabited the land where they live. As a result of deforestation in this area, flooding of rivers and washing out of bridges has become more common. The regional climate may be changing as well. Hunting is now rare within a large radius of the nearby city. There will be no place to take adults or children to demonstrate how the forest provides habitat, mitigates flooding, and affects the climate. Each rancher in the area has, individually, chosen money over nonmonetary resources for the future. A few of the children may inherit fat bank accounts and big ranch houses, but the landscape and most of the population will be impoverished for many decades to come.

In the second example, oil delivered in the future is worth less than oil delivered now, which has profound implications for the way we value future resources. In effect, we are saying that natural resources used in the future are worth less than those used now. If future oil has discounted value, then there is little incentive for you or me or anybody else to conserve oil. If next year's oil is worth only $95 per 100 gallons, and oil delivered several years from now is worth even less, then it is harder to justify spending money now on better insulation or other improvements in energy use efficiency that would conserve future oil that has so little value. Assuming that this discounting of the future is correct, we are better off spending money on oil now and investing the money that we might have spent on energy conservation.

Not only will this disincentive to conserve oil mean that future generations will have less oil available, but as we consume the oil, we also will be releasing carbon dioxide into the atmosphere, which will contribute to global warming that future generations must endure (which is the subject of the next chapter). When the future value of something is discounted, be it timber, oil, water, air, or any consumer product that depends upon natural resources for its production, the implicit assumption is that natural resources of the future have less value than natural resources of the present.

What's Wrong with This Picture?

Any parent or grandparent knows that there is something wrong with this notion that the future is worth less than the present. After all, we are willing to make sacrifices for the sake of our children's future: We work hard, save for their college educations, leave most of the money we have to our children when we die, and generally strive to make a better world for our children than the one we inherited. Why then, does economics tell us that the future value of the resources that our children will depend upon for their well-being should be discounted? Something does not make sense.

One response that I have heard some economists make is that discounting the future is simply a reflection of our natural tendency to be impatient. Despite the savings and sacrifices that many people make for their children's future, most of us do not save as much as we should because we are impatient in our desire to consume and enjoy material goods now. "Buy now, pay later" is an effective marketing slogan because it works. Large credit card debts, car loans, and mortgages provide convincing evidence. In truth, many of us know in our hearts that saving for the future would be prudent, but we yield to temptation to spend now. Likewise, we know that protecting natural resources for the benefit of future generations is the right thing to do, but we often yield to the temptation of ignoring the well-being of future generations for the sake of more consumption in the short-term.

I reject, however, the justification of discounting used in economics as simply an unavoidable reflection of human frailty. It is true that discounting may reflect the frailty of impatience to consume now, but we need not accept it as unavoidable. We also tend to drive our cars too fast, because we are in a hurry, enjoy the thrill of driving fast, or want to show off to someone. Despite these human frailties, we have laws against speeding that are pretty well enforced with large fines, because we know that speeding is dangerous for both the speeder and the other occupants of the road.

Likewise, disregard for the environment is unhealthy for the exploiter and for the other occupants of the environment. Like speeding laws, we have laws and policies that influence the way people make money, borrow money, and spend money. Some ways of making money are illegal, despite a natural human tendency for greed. Interest rates, and hence discount rates, are influenced by government policy for a variety of reasons. Income taxes penalize income earned on taxable investments, reward income from tax-free investments, reward charitable giving, reward borrowing money for tax-deductible mortgages, and are neutral for nondeductible interest payments. So if we want to encourage people to drive their cars responsibly and to earn, borrow, consume, invest, and save money in responsible ways, we have the tools of regulation, interest rates, and tax incentives to influence those behaviors. Discounting the future is not an inevitable outcome of human frailty that must be tolerated, any more than reckless driving is. Rather, deciding whether discounting should be used in cost-benefit analyses, and if so, at what rate, should be a deliberate decision.

FINDING THE RIGHT DISCOUNT RATES

One group of economists, including some of the new ecological economists, argue that the problem is finding the *right* discount rate. Usually, the lower the discount rate, the greater the value placed on future resources. In the example of timber on the back forty acres, it would make sense not to cut if the discount rate were only 4 percent, instead of the 8 percent assumed in the example given. We could afford to rent our patch of Brazilian rain forest for research for a longer period of time if the rancher used a lower discount rate to calculate the rent. Some economists argue that the future environment would be valued more (or at least devalued less), if lower discount rates were used in cost-benefit analyses.

However, there are exceptions. Dams are built in order to provide electricity, irrigation water, drinking water, flood control, and recreational boating in the future. The value of those resources delivered in the future during the lifetime of the dam must be calculated using an appropriate discount rate. The dam will also have some negative impacts on the environment, such as flooding the valley upstream with a huge reservoir and stealing the water from the river downstream that would normally help keep fish populations healthy. In this case, if a high discount rate is used to calculate the value of the future irrigation water provided by the dam, then the monetary benefits of the dam will be less, and saving the river just might come out on top in a cost-benefit analysis. So finding the *right* discount rate is tricky, and it can have enormous consequences for calculations of costs and benefits of any project or activity that affects the environment.

The discount rate is normally based on a *reasonable rate of return,* an estimate of what one could reasonably expect as a return on an investment. The word reasonable leaves a lot of wiggle room and provides an opportunity to skew the analysis in whatever direction is desired. As in the example I gave for the purchase of oil, the rate of return earned on an investment depends on the level of risk that the investor is willing to accept. High-risk investments may yield high returns, whereas low-risk investments are usually fixed at low rates of return.

Moreover, people differ in their degree of impatience and risk taking. Economists refer to these choices as *personal discount rates,* which vary from person to person. Some people want their money now so that they can spend it on the consumption of material goods now (high personal discount rate), whereas others are willing to save and invest for a long-term payback and material comfort in the future (low personal discount rate). Similarly, societies, cultures, and governments can be characterized by generalizations about their typical discount rates. Americans tend to be spenders (high discount rates, large national debt), and Japanese tend to be savers (low discount rates, big buyers of U.S. government bonds).

Personal discount rates are derived from personal decisions, often made impulsively with little reflection on future consequences. Corporations and governments, on the other hand, are more deliberate about choosing discount rates for their cost-benefit analyses, although the assumed discount rates may not be examined as carefully as they should be. Given the wide range of possible returns on investment, there is considerable uncertainty in the selection of the *right* discount rate, and hence, there must also be uncertainty in the results of the economic analyses.

THE PERVASIVENESS OF DISCOUNTING

As long as the economy is growing and the GNP is increasing, the rate of return will be positive, and as long as you can make money on monetary investments, then the value of the future environment will be discounted according to current neoclassical economics. In the industrialized world, we have a track record of usually generating economic growth. Hence, our legacy has been to use resources rather than conserve them and to pollute resources rather than safeguard them.

By discounting the future, we can rationalize away the need to spend money now to prevent pollution. A recent cost-benefit analysis of the actions needed to avoid global warming came to the conclusion that it would be best to wait before adopting most proposed actions to reduce greenhouse gas emissions, such as investing in finding alternatives to our dependence on gas, coal, and oil. It will cost money to accelerate the development and use of more efficient engines and more efficient heating and cooling, or to substitute solar power and other renewable energy sources for fossil fuels. This argument for delaying is based on two assumptions: (1) the money we will save from postponing this research, development, and implementation of alternative energy sources can be invested where it will earn a positive return; and (2) technological developments will occur in the meantime that will make it easier and cheaper in the future to improve energy efficiencies and to phase out fossil fuels. The first assumption is the classic discounting of future resources that we just discussed, and it assumes that we do not value the natural resources that we leave to our children and grandchildren. The second assumption is the same "Custer's Folly" fallacy that I discussed in Chapter 1—that technological developments will rescue future generations from our current unsustainable habits, allowing our grandchildren to prevent further global warming and to tolerate the climate changes that we have already caused. Clearly, these are debatable assumptions, and it would be prudent to temper these optimistic assumptions with precaution.

THE PRECAUTIONARY PRINCIPLE AND ECONOMIC THEORY

If we value our children's future options to use natural resources more than the discount rate implies, and if we are more cautious than Custer, then our choice will be to accept many of the costs now of avoiding soil erosion, global warming, groundwater pollution, and vast destruction of forest habitat. We accept these present costs for the sake of our children's future well-being. In the jargon of policymakers, this avoidance of future risks is called the *precautionary principle.*

Unfortunately, the precautionary principle often is not followed, because our natural parental instincts to be cautious are at odds with the way that economics is used to justify policy decisions. Instead of starting with a vision of the world we want to leave to future generations and then allowing economics to help us find the most efficient way to achieve that goal, we are letting the economic tail wag the dog by misusing the tools of economics to define the goals. Until very recently, the neoclassical economists who do the cost-benefit analyses have been assigning all of the discount rates. Not surprisingly, when the chosen discount rates are high, the future values of our natural resources are discounted, and then our policymakers use these economic analyses to justify policies that continue to overexploit and pollute those natural resources.

Some economists have argued that the discount rates for individuals making market-driven decisions should be different from the discount rates used for calculating the social costs of governmental policy decisions. The Brazilian rancher, who was concerned about amassing financial assets that he could use now or pass on to his children, insisted on immediate payback (rent) for the delay in cutting his trees. He used a high discount rate based on fairly aggressive investment options available to him in the marketplace. It would be hard (and in our case, was impossible) to convince the individual rancher to use a lower discount rate for his personal decisionmaking about his land and his own narrow self-interest. In contrast, local, state, and national governments adopt regulations, tax policies, and credit policies that affect the behavior of thousands of ranchers and that have environmental, economic, and social costs and benefits for large regions and populations. Some economists argue that governments should use a discount rate that is lower than prevailing market discount rates when considering the collective environmental and social costs and benefits of their policy options. The lower discount rate would usually give more weight to environmental concerns and to the costs and benefits to future generations. Other economists argue that the discount rate is the wrong place to make these adjustments. Instead, they argue that more effort needs to go into improving the accuracy of the cost-benefit analyses by finding the "right prices" for environmental goods and services (as discussed in Chapter 3).

In any case, cost-benefit analyses are no more correct than their assumptions, including both the prices of ecosystem services and the assumed discount rate, with its implied devaluation of future natural resources. Although objective comparisons of monetary trade-offs in cost-benefit analyses are worthwhile and often enlightening exercises, these analyses are never free of subjectivity or of underlying value systems, nor should they be. Our moral and social values must be considered along with the economic factors.

One of the first lessons of neoclassical economics, which is often subsequently forgotten, is that it says nothing about what kind of distribution of land and wealth is

morally right or socially acceptable. Many economists con-
duct research that reveals the consequences of various systems of distributing wealth, but neoclassical economics
does not indicate whether it is good or bad to have a small
elite upper class with a huge poor underclass, or a completely egalitarian distribution of wealth, or something in
between with a large middle class. How wealth and land
ownership *should* be distributed and how much poverty
should be tolerated are determined largely by social struc-
tures and government actions. Neoclassical economic theory explains only how the market will respond, given an
agreed upon initial set of conditions that define the distribution of land and wealth.

The same applies to the distribution of wealth and land
ownership among generations. We should not expect neoclassical economic theory to tell us how much of our natural resource wealth we *should* leave intact for the use of
future generations. Our decisions about distributing the
wealth we obtain from natural resources among present
and future generations are best made according to what
we think is right and wrong, not according to the prediction of an econometric model. Once a decision has been
made as to how much we want future generations to inherit, both in terms of still untapped natural resource
wealth and in terms of polluted and degraded resources
that are analogous to debts, then economics can help us
find the most efficient ways for the market system to
achieve those chosen objectives.

At least that is the way it should be. Unfortunately, instead, we have too often allowed neoclassical economic
analyses to justify our current uses and abuses of natural
resources. The calculated benefits to future generations

> If, however, economic ambitions are good servants, they are bad masters.
> —**R. H. Tawney, *Religion and the Rise of Capitalism***

often do not appear to justify the costs of changing our current generation's wasteful and abusive habits. Perhaps the economists, themselves, are not to blame, although neoclassical economists have certainly provided the economic tools that have been used so effectively by greedy pursuers of short-term profit. Every society will have greedy people who will seek gains for themselves at the expense of others, both present and unborn. To be sure, there is a fine line between this destructive greediness and the entrepreneurial self-interest motivation that can also have positive effects for society at large. The point, however, is that the drive for short-term profit must be kept in check by an economic system that is designed to value the future at least as much as the present. Yet our current system, which relies so heavily on discounting in economic analyses, condones short-term profit taking that infringes upon our precautionary values of leaving a rich natural inheritance to our children.

Ecological economics is making progress in addressing the important details of prices and discount rates, but economic analyses alone cannot provide the basis for the types of value judgments that are needed. Perhaps I would feel better off now if I spent money on pleasurable things that bring immediate gratification, rather than saving for my child's college education or for my retirement. I resist that temptation, however, because, discount rates aside, I put value on my future and my child's future. Economic analyses help me determine how much I must save now in order to afford a community college, state university, or private college or university for my child and how much I have to save now in order to ensure various degrees of comfort in retirement. Even with the most expert professional financial planning advice, however, only my spouse and I can decide which of these options is right for our family. Similarly, economic analyses can provide helpful information on what it will cost to protect and properly manage the environment now for the sake of the quality and abundance of natural resources inherited by our children. We may fine-tune our vision of the future according to these cost estimates, but our decisions must be based principally on our values of right and wrong, not upon economic calculations of a discounted future.

Why do most of us apply the precautionary principle at home, yet allow our governments to ignore this principle? The precautionary principle is instinctive, and I trust it to be right. But most of us, myself included, have been ignorant about the devilish little secret of discounting that pervades our economic and policymaking systems. Ecological economics may or may not be able to fix the underlying flaws in our current economic approach to discounting the future. Economists of all kinds are debating among themselves how best to set discount rates. They are debating whether adjusting discount rates is an effective tool for protecting the environment, or if variable discount rates adopted by governments might backfire by creating unforeseen perverse economic incentives. I do not claim to know the answer. I do believe, however, that bringing the importance of discounting out in the open and showing how the assumed discount rate affects the outcome of cost-benefit analyses will help make these analyses more transparent and understandable. We must weave together our instinctive parental precaution with transparent economic analyses as we make decisions that affect the environmental inheritance passed on to the next generation.

5 🍃

Internalizing the Externalities

Buying a Bunch of Blue Sky to Limit Global Warming

I FIRST HEARD THE EXPRESSION, "buying a bunch of blue sky," from a lawyer who told me about advising a doctor who was considering buying into an existing medical practice. After reviewing the details of the contract, the attorney advised his client that the medical practice did not have many assets and that paying a significant sum of money to join a medical practice with few tangible monetary assets was like "buying a bunch of blue sky." The doctor was unmoved by this advice, because, in this case, he knew that buying "blue sky" would entitle him to a profit of many green dollars. This use of blue sky as a metaphor for something without much value left a big impression on me, however, because it was at such odds both with my childhood impressions and what I have since learned about blue sky.

LITTLE BIG SKY

Pass through the Big Sky Country of my childhood home state of Montana and you will be awed by the vastness of the land and the blue sky above it. Most of my respect for nature came from the smallness I felt while hiking as a teenager in the vast wilderness of the Montana mountains under its big blue sky. There was no question in my mind that the sky was what made Montana so special. The Big Sky over Montana and every place else on the planet, however, now contains extra carbon dioxide, traces of metals, radioactive fallout, and everything else that we have thrown up into the air from our smoke stacks, tailpipes, and nuclear bombs. Although a few places like the Montana "wilderness" appear to be less affected than others, no place on earth, however remote, remains pristine or immune from human impact. Now that I am a scientist studying how the ecology of the earth is linked to its atmosphere and how humans have changed the atmosphere, the Big Sky of my childhood has become rather small relative to our technological ability to modify it. Although the size of the sky seems to have shrunk in terms of how vulnerable it is to being changed by humans, I still hold to my childhood instinct that blue sky has value.

THE VALUE OF BLUE SKY

Supporters of the University of North Carolina Tarheel basketball team, with its sky-blue school colors, are convinced that the sky is light blue because God is a Tarheel. Having gone to school at rival North Carolina State University, I prefer a more scientific explanation. The sky is blue because tiny particles in the atmosphere cause waves of light from the sun to bend (diffract) around them, which happens to yield a blue color. When water vapor diffracts light, we see beautiful rainbows, but most of the diffraction of the sun's light just gives us a pretty blue sky. Aside from being pretty, the sky's blue color shows that the atmosphere is not simply a void. There are things up there in the sky, such as gases and tiny particles called aerosols, that bend light, reflect light, absorb energy, and trap heat like a blanket around the earth. That might be hard to believe on a windy winter day, but it is absolutely true. Water vapor, carbon dioxide, methane, and other gases help trap some of the sun's energy inside the atmosphere. The earth stays much warmer than it would if those gases were not there. This heat-trapping effect has been called the *greenhouse effect* because it behaves like the glass of a greenhouse, which traps the sun's energy within, warming the greenhouse. A more common experience is getting into a scorching hot car that has been parked in the sun—the sun's energy enters through the windshield, but it cannot get back out and thus builds up inside the car. Perhaps the greenhouse effect would be more widely understood if it were dubbed the *parked car effect,* but I suppose the latter does not quite have the same poetic ring.

Although the popular press often refers to the greenhouse effect as a *theory,* it is no more questionable than the theory of gravity or than the theory of parked car effect. If it were not for the existence of heat-trapping gases in our atmosphere, the earth would be a frozen ball of ice, and life as we know it would not exist. Concerns about *global warming* stem from evidence that the greenhouse effect is growing stronger because we hu-

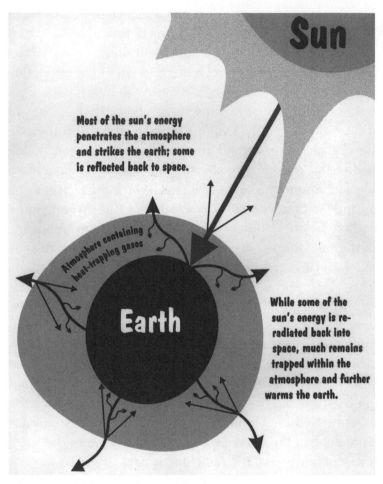

Most of the sun's energy penetrates the atmosphere and strikes the earth; some is reflected back to space.

Sun

Atmosphere containing heat-trapping gases

Earth

While some of the sun's energy is re-radiated back into space, much remains trapped within the atmosphere and further warms the earth.

Illustration of greenhouse effect

mans are dumping more heat-trapping gases into the atmosphere as we burn gas and oil in our cars, furnaces, and power plants and as we cut down and burn forests. Some aspects of the global warming issue are being debated, but no one challenges the fact that by burning coal and oil and clearing forests, we humans have increased

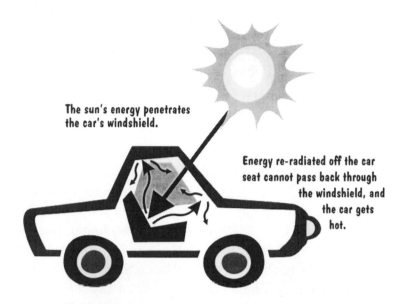

The sun's energy penetrates the car's windshield.

Energy re-radiated off the car seat cannot pass back through the windshield, and the car gets hot.

Illustration of parked car effect

the concentration of carbon dioxide in the atmosphere by about 25 percent. If business goes on as usual, we will double atmospheric carbon dioxide by the end of the twenty-first century.

Global warming has been the subject of study by an international commission called the Intergovernmental Panel on Climate Change (IPCC). Working since the late 1980s, the IPCC has accomplished an unprecedented and remarkable feat: engaging about 500 scientists as coauthors of reports and another 500 scientists as reviewers representing more than seventy countries. A consensus has recently emerged among these scientists that the expected warming is strong enough and fast enough to have important consequences to the global environment and economy. Although many details about the warming scenario are still being debated among scientists, the IPCC has shown that the earth has warmed a little more than one-half degree centigrade (about one degree Fahrenheit) during the past century. This distinguished group of climatologists, oceanographers, geochemists, and ecologists were able to reach the following consensus statements: "The observed warming is unlikely to be entirely natural in origin." "The balance of evidence suggests that there is a discernable human influence on global climate."

After examining a range of plausible estimates for future warming, including the lowest and highest rates of warming that are consistent with our understanding of the climate system and how it is being altered by humans, the IPCC goes on to conclude: "In all cases the average rate of warming would probably be greater than any seen in the last 10,000 years."

As responsible scientists always do, the IPCC report couches these strong statements with appropriate caveats about uncertainties in the data sets and models used to reach this conclusion. Nevertheless, a "fingerprint" of human activity on climate was clearly discerned from the pattern of when and where warming has been observed. In conclusion, the warming has been reliably detected, it is most likely caused by us, and more warming is almost certainly on the way at a rate that is unprecedented since humans first developed agriculture.

I am not describing changes that will happen hundreds of years from now, but rather only tens of years into the future and well within the lifetimes of the majority of people alive today. We have already observed some of these changes, and convincing signs of global warming are found all over the world. The sea level has risen between four and ten inches during the past one hundred years. Arctic sea ice is melting and glaciers are receding throughout the world. The migrating patterns of birds have changed. Coral reefs are suffering from unusually warm seawater that makes them susceptible to disease. Cloudiness has increased and summer nights are not cooling down as much as normally expected. Autumn frosts are late, spring is often early, and unusual weather patterns are becoming common, including droughts, warm spells, and "freak" storms that bring unusually large amounts of rain or snow. The warmest years in recorded history have been 1995, 1997, 1998, and 1999. Taken together, the preponderance of evidence makes a strong case that global warming is well underway.

We cannot predict with certainty how fast the earth will continue to warm, but our best estimate is that it will warm, on average, about four degrees Fahrenheit during the next one hundred years. A change of four degrees Fahrenheit is nearly half as much as the amount of warming that has occurred since the last ice age, about ten thousand years ago. Instead of having thousands of years to adjust to this change, humans, plants, and animals will have only one hundred years to attempt to adapt to the next warming of four or more degrees. Within the next fifty years, which is well within the lifetime of my son, the sea level is expected to rise another eight inches. These are large and rapid changes relative to natural changes in climate that have been recorded in historical and geological records.

Of course, there is a dissenting opinion among a small group of scientists, most of whom are generously supported by the coal and oil industry. These naysayers of global warming are addressed later in this chapter. For now, let us note that blue sky must have value because it helps keep the earth's climate hospitable, and let us concentrate on the ecological and economic consequences of global warming.

THE VALUE OF A HOSPITABLE CLIMATE, OR WHAT'S SO BAD ABOUT WARMER WEATHER?

Does it really matter if global warming is for real? After all, milder winters do not sound all that bad. Unfortunately, just as our modern way of life removes us from the sources of our food, water, fiber, and the biophysical basis of our well-being, it also tends to make us think of climate as simply a matter of what type of coat we need to wear. By being several steps removed from our dependence on natural resources, most people do not recognize that changes in climate cause profound disruption of ecological, economic, and social systems.

In 1997, over 2,500 scientists from the disciplines of climatology, geochemistry, atmospheric chemistry, biology, ecology, forestry, agriculture, medicine, and others signed a Scientists' Statement on Global Climatic Disruption. Referring to the IPCC studies mentioned above, these scientists concluded:

> We endorse those [IPCC] reports and observe that the further accumulation of greenhouse gases commits the earth irreversibly to further global climatic change and consequent ecological, economic and social disruption. The risks associated with such

changes justify preventive action through reductions in emissions of greenhouse gases. . . .

The warming is expected to expand the geographical ranges of malaria and dengue fever and to open large new areas to other human diseases and plant and animal pests. Effects of the disruption of climate are sufficiently complicated that it is appropriate to assume there will be effects not now anticipated. . . .

Our familiarity with the scale, severity, and costs to human welfare of the disruptions that the climatic changes threaten leads us to introduce this note of urgency and to call for early domestic action to reduce U.S. emissions via the most cost-effective means. We encourage other nations to join in similar actions.

This concern is shared by many economists. A group of over two thousand economists also issued a statement in 1997, registering their concern about the significant risks posed by the scientific findings of the IPCC: "As economists, we believe that global climate change carries with it significant environmental, economic, social, and geopolitical risks, and that preventive steps are justified."

Unfortunately, these statements by experts in ecology, climatology, medicine, economics, and other specialties may not engender enough concern among the general public, because many of the risks of climate change are hard to perceive in everyday life. The immediate direct effects of global warming on people are serious enough: the spread of tropical diseases, summer heat waves, flooding near coastal areas as the oceans rise a foot or so, and perhaps more frequent hurricanes. More importantly, however, the effect of climate goes beyond the inconveniences of sickness, heat waves, storms, floods, and other unusual weather, as well as beyond the conveniences of fewer days with winter coats. Remember that we depend upon soils for producing food and upon forests for providing wood, clean water, and habitat for plants and animals. For our soils to remain productive and for our forests to remain healthy, we must take a lesson from Goldilocks—the climate should be "not too hot" and "not too cold" but "just right."

Every place on earth, the people, plants, animals, and even the microbes have adapted to the climate in which they live. What an Eskimo Goldilocks would consider "just right" is far cooler than what a Hawaiian Goldilocks would prefer. To some extent, people can pick up and move to find the climate that suits them best, although human migrations have their hardships too. The trees and the other plants can also migrate by gradually invading new territory when their seeds are spread by wind and by birds to a new habitat that has a favorable climate for that particular type of tree. In the time frame of hundreds and thousands of years, forests can adapt to changing climate. Most of the temperate forests of North America now stand where there were only giant glaciers and bare rock ten thousand years ago. The ancestors of these modern trees must have lived further south ten thousand years ago when the temperate climate that they needed was also further south. As the ice age ended, these trees were able to migrate north gradually, probably at the rate of 5 to 30 miles per century, during several thousand years as the glaciers retreated.

In contrast to this slow pace of natural migration of forests, the warming caused by our recent dependence on coal and oil burning is happening much too rapidly for forests to adapt. In order to keep up with the warming that is most likely to occur during the next one hundred years, the plants would need to migrate northward about 120–180 miles per century, or one to two miles every year. This rate of migration is much faster than what has occurred naturally since the last ice age and is probably not possible.

As the earth warms, many of the existing forest trees will become increasingly maladapted to the new climate in which they find themselves, and they will become susceptible to disease and fire. In some cases, it may take several decades or more for the adult trees to succumb to death. New vegetation will take the place of dying trees, but our understanding of plant succession suggests that the new vegetation is most likely to be made up of "weedy" species of plants that are better adapted to disturbed habitats than are the slow-growing sequoia, cedar, hemlock, and beech. If and when the climate is stabilized again, new forests eventually will become established according to the new distribution of climatic zones. Our present forests as we know them, however, will go through a transition of many decades that could be very destructive. If the warming is at the higher end of the range predicted by climatologists, then the habitat for many forest trees could move northward by as much as 350 miles, thus requiring many decades or centuries for the trees to migrate and for the new forests to become established. As this is uncharted territory, I cannot assign a probability to this scenario, other than to say that the evidence in support of this large-scale destruction and redistribution of forests is as good or better than the evidence against it.

Neoclassical economists calculate that the effect of losing forests on the GNP is negligible—logging of native forests is already on the decline in the United States, and like agriculture, it is only a small fraction of the nation's GNP. The biggest and most important effects of the loss of forests will be the loss of good clean water that forests help purify, loss of the heat-absorbing capacity of the forest that helps regulate the climate, and loss of habitat for a wide variety of plants and animals, but those effects are not currently included in the GNP.

Technology is unlikely to find substitutes for these essential services provided by forests. Technologists have dreamed up expensive schemes to purify enormous quantities of water pumped from streams and from groundwater in an attempt to replace the good-quality water now provided for free by forest watersheds. Giant space shields positioned in orbit over the earth and other extravagant schemes have been proposed to replace the beneficial effects that forests have on climate. These ideas are silly not so much because they are technologically infeasible—perhaps they could be made to work a few decades from now—but because they attempt to replace something that already works so well. Simply keep the climate from changing rapidly and keep the forests in good health, and we will have a proven natural "technology" that we know will provide what we need. Start tinkering by replacing forests with new, unproven technologies, and we take a giant risk that is unnecessary and imprudent.

It might seem that we could adapt our agricultural systems to a changing climate more readily than our forested ecosystems. We can add more irrigation, for example, in those farming regions like Iowa, where summers are expected to get hotter and drier. Changes in rainfall are more difficult to predict than temperature changes, but there is good agreement that coastal areas generally will receive more winter rain, while midcontinental areas, where the most productive agriculture occurs, generally will become drier in the summer. As shown in Chapter 6, however, increasing irrigation will cause another, bigger problem by further reducing already depleted, and in some cases polluted, groundwater.

Maybe the cornfields of Iowa could be moved north to Minnesota when Iowa becomes too hot and dry. Similarly, Canada may gain agricultural land as the climate warms and the United States loses land with suitable climate for agriculture. The assumption that productive agriculture can simply shift northward may have some basis, but it is dangerously simplistic. The northern limit of wheat farming in Canada, for example, is determined by both climate and geology. Even if the climate becomes favorable for growing wheat further north than is possible today, most of the types of soils that have formed over the geologic formations of northern Canada, such as the Canadian Shield, are not like the naturally fertile prairie soils found further south. The fertilizer inputs that would be required to move productive agriculture north to these naturally infertile soils in order to follow a changing climate would be economically impractical. Our agricultural systems are highly productive where they are under the current climate, although they do suffer from some significant problems, such as polluted groundwater and soil erosion. Changing the climate and trying to move this productive agriculture of the midwestern United States further north would create new problems and more risks.

Driving through the tall fields of Iowa corn in the late summer or through the brilliant New England forest on an autumn day, it may seem like they are in good shape and that the threat of global warming is remote and abstract. A person with high blood pressure may also appear healthy and chipper, but we trust medical science when it puts greater faith in a blood pressure reading than the patient's cheery demeanor. To be prudent, people with high blood pressure moderate their diet and change their exercise habits. Similarly, the climatologists, geochemists, and ecologists have measured the temperature of the earth and have shown that it is becoming warmer at an unnaturally rapid rate. We know that increasing global temperature is a strong indicator of trouble to come, even if the symptoms have not yet overcome us. To be prudent, we must heed this warning by moderating our diet of fossil fuels and by exercising better judgment on how we manage the invaluable resource of the atmosphere and the hospitable climate it provides.

CAN THE INVALUABLE ATMOSPHERE BE GIVEN MARKET VALUE?

In one sense, the value of blue sky becomes most obvious when polluted sky becomes more visible. When the air pollution hanging over a city can be seen as ugly smoke or haze, when it can be smelled as choking odors, when children must play indoors during bad pollution days, and when more people are treated in hospitals for respiratory conditions, then the value of unpolluted air becomes obvious, at least in a general, qualitative way. Some economists have tried to make specific dollar-for-dollar comparisons of the costs of cleaning up air pollution to the cost of hospital treatment and days lost from work due to air pollution. As discussed in Chapter 3, these cost-benefit analyses are rife with uncertain assumptions about how air pollution does or does not interfere with economic productivity. They also miss many of the nonmonetary benefits of clean air, such as kids burning off energy safely while playing out-doors and all of us enjoying taking a good deep breath without fear of inhaling toxic air pollutants.

The value of clear blue sky is more difficult to assess when the form of air pollution is less visible and odorless and when the effects are dispersed over public forests and lakes. So-called *acid rain* is a form of air pollution caused by nitrogen oxides and sulfur oxides emitted from smoke-stacks and tailpipes. The pollutants are carried downwind and then deposited back to the ground in rainwater. The pollutants make the rainwater more acidic, and the rain acidifies lakes and soils. In many cases, excess nitrogen and sulfur nutrients from the rainwater also alter the types of plants and animals that can grow in the lakes and soils. Once it is carried downwind and mixed with the atmo-sphere, this form of pollution cannot be seen or smelled. More importantly, the cause, such as industrial emissions in midwestern states like Ohio, can be a long distance from the effect, such as the downwind forests and lakes of New York, New England, Ontario, and Quebec that receive the acidified rain. Not surprisingly, the downwind governor and citizens of New York state have a different view of the value of clean air than do the upwind governor and some of the citizens of Ohio. Their views about air pollution usu-ally have more to do with whether they live upwind or downwind rather than their political party affiliation or liberal-conservative leanings. Similarly, air does not need a passport to cross international boundaries, nor can the pollution be stopped at the border by customs inspectors. Hence, Canadians and Americans often have different perspectives on the value of clean air and the costs of cleaning it up.

Global warming brings the air pollution issue to the global scale, and the offending heat-trapping gases that circle the earth cannot be seen or smelled. We must rely upon experts with sophisticated instruments and computer models to describe these risks across the globe, rather than rely on our own visual and olfactory senses and common sense at home. Up to now, the industrialized countries of North America and Europe have caused most of the problem, having used most of the coal and oil that produces carbon dioxide and other greenhouse gases. Before long, however, China could overtake the United States as the biggest contributor to greenhouse gas emissions. The cutting of forests in tropical countries like Brazil, Indonesia, Malaysia, and Congo is also an increasingly important cause of accumulation of heat-trapping gases in the atmosphere. The stakeholders in the contentious global warming debate include poor and rich countries, coal companies and environmental organizations, and most importantly, citizens with concerns of economic well-being, health, and quality of life. This broad array of parties with different value systems ensures that the task of assessing the value of blue sky and the hospitable climate that it provides will be challenging, to say the least.

So what is a bunch of blue sky really worth? That is probably an unanswerable question, because, like soil, the atmosphere is one of those *in*valuable and nonsubstitutable resources. We can, however, analyze the claims that air pollution is an unavoidable consequence of economic prosperity or that it would be too expensive to avoid. Every time that an industry group has warned that a particular measure to reduce air pollution would cause economic ruin, the doomsday prediction has turned out to be false once pollution abatement policy was in place. Consider the following examples:

Taking the Lead Out of Gasoline. This was supposed to drive up gasoline costs and cripple the transportation industry and all industries that depended upon transportation. Instead, gasoline is now cheaper, after correcting for inflation, than when it contained lead. Lead contamination caused by polluted air coming from burning leaded gasoline has decreased, and fewer cases of neurologic disorders in children caused by lead poisoning have been documented. Where are the intolerable costs?

Protecting the Ozone Layer. A growing "hole" in the ozone layer has been conclusively shown to be caused by man-made chemicals called chlorofluorocarbons (CFCs), which are used as refrigerants for air conditioning and as solvents in the computer industry. Unfortunately, the refrigerants are eventually vented to the atmosphere, and the cleaning solvents evaporate, causing the CFCs to accumulate in the atmosphere. This form of air pollution destroys the protective ozone layer that naturally exists high in the atmosphere. The ozone normally screens out harmful ultraviolet radiation from the sun, but as this ozone is destroyed, we are exposed to higher doses of cancer-causing radiation. Many plants are also damaged when they receive too much ultraviolet radiation, and that can affect the amount of food available to animals of all kinds, including humans. To avoid these threats, CFCs have been banned, but the banning was initially opposed because of predictions that it would wreak economic havoc. The ban is now in place, other chemicals have been developed to replace the CFCs as cleaning solvents in the computer industry and as refrigerants, the price of computers continues to drop, and air conditioning is still affordable. Concentrations of ozone-eating CFCs in the atmosphere are declining, but the predicted economic upheaval barely made a blip on the economic screen.

Reducing Sulfur Emissions That Cause Acid Rain. As already explained, sulfur is one of the gases emitted from tailpipes and smokestacks that results in acidified rain downwind, which harms lakes and forests and can even damage buildings and statues. As the result of a novel policy that used market forces to give industry incentives to reduce air pollution, the costs of achieving significant reductions in sulfur emissions have been about one-tenth of what industry groups predicted. We have not solved the acid rain problem yet, because we need to make similar reductions in nitrogen oxide emissions (another kind of gas that produces acid rain), but the predicted severe economic hardships resulting from sulfur reduction did not materialize.

INTERNALIZING AN AIR POLLUTION EXTERNALITY

This last example of reducing sulfur emissions is particularly relevant to global warming, because a very similar strategy has been proposed to reduce greenhouse gas emissions. We would be wise to take a lesson from the novel, market-driven strategy that has made both environmental and industry groups relatively content.

Although the atmosphere is an invaluable resource, a marketplace value has been constructed for blue sky for the sake of protecting it. An interesting success story of ecological economics has been the introduction of permits for smokestack emissions of sulfur gases that can be traded for money. The U.S. government has decided to accept a certain level of sulfur air pollution, and it issues permits to polluters for emissions of pollutants that can be traded like commodities. So if DoGood & Co. have found a clever way to cheaply reduce their smokestack emissions of sulfur below the level of their pollution permits, they can sell their unused leftover pollution permits to Evil Inc., which still finds it cheaper to buy additional pollution permits than to clean up its act. DoGood & Co. realize an extra profit from the sale of their pollution permits. Evil Inc. has not been forced to clean up, but they pay extra for the right to pollute. The cost of air pollution will be passed on to the consumers of Evil Inc.'s products in the form of higher prices, but DoGood & Co. may either lower their prices or pass on larger dividends to their stockholders. Clean blue sky has been given a market value, it is traded within the marketplace, and its monetary value even gets added to the GNP! A bunch of blue sky can be sold for a considerable sum of money, contributing to black ink on the bottom line!

To be precise, it is polluted air that is traded within the marketplace, not blue sky. This system of tradable pollution permits was developed, however, because blue sky was valued at least as much as the sum total of all traded pollution permits. Clearly, this is only a partial monetary valuation of blue sky, but it is one that helped get an important job done—reducing one of the many forms of air pollution.

Air pollution, like most forms of pollution, has been dubbed an *externality* by neoclassical economists, because its costs are usually external to the market economy upon which neoclassical economics is based. Anything that is not traded efficiently within the marketplace is external to the economic pyramid, and it is often ignored in neoclassical economic analyses. Some economists study these externalities as an example of what they call *market failures*. Tradable pollution permits have *internalized* this externality by making air pollution into a commodity that can be traded quite successfully and effectively within the marketplace, thus correcting the market failure. The pollution that harms the ecological system and was previously unaccounted for in the neoclassical economic system now has a monetary value that the economic system can grasp. Most importantly, this economic handle provides a clever tool to reduce the pollution problem.

Assuming that the government can make sure that everyone plays by the rules and pollutes no more than allowed by the pollution permits that they have bought, then both cleanliness and entrepreneurial innovation are rewarded, whereas dirtiness and stagnation are penalized by market forces. The key to making this scheme work to reduce pollution in the long term is that the government must gradually issue fewer and fewer pollution permits, so that the permits become more scarce and more valuable. Eventually, even Evil Inc. will find innovative ways to adapt to this new market pressure and reduce its pollution.

In this example, government has played a key role by simply decreeing that it will tolerate only a certain amount of pollution, but that the free market can sort out the most economically efficient way of determining who does the polluting and who cleans up first. Rigid governmental regulations and prescriptions that often miss their mark have been avoided, while marketplace forces spur innovations that ultimately reduce the costs of compliance with pollution limits. Although this method relies upon market forces for its success, it certainly is not laissez-faire economics, because government must first create the demand for pollution permits before the market forces can trade them.

This system has worked within the United States for reducing regional sulfur pollution problems. For problems that do not respect international borders, like global warming, this approach may be more difficult to implement, because it will be a tremendous challenge to determine how many pollution permits each country should start out with. Should the United States and European countries get more greenhouse gas emission permits because they already have more smokestacks and tailpipes, or should China and India get more permits because they have more people? Moreover, it will be very difficult to monitor greenhouse gas emissions and to confirm compliance with traded permits because there are so many different sources of greenhouse gases, from smokestacks to landfills to fertilized croplands. Economists correctly argue that an energy tax or a carbon tax would be easier to administer and would be at least as effective, if not more effective, than tradable permits for greenhouse gas emissions. Tax increases of any kind, however, even when offset by reductions of other types of taxes, are so politically unpopular that tradable emission permits presently seem to be one of the few viable approaches for creating economic incentives to reduce greenhouse gas emissions.

TURNING THE CORNER ON THE CARBON ECONOMY

Representatives of the nations of the world made a modest step in the right direction at the culmination of an international negotiating session in Kyoto, Japan, in December 1997. It was agreed that most of the industrialized nations would curb their emissions of greenhouse gases, so that by the year 2010, their emissions would be 7–8 percent lower than they were in the year 1990. Scientists have since pointed out that this size of a cut will make only a tiny dent in the problem. Environmentalists wanted deeper cuts in emissions, while the oil and coal industry lobbied hard for no cuts at all. As usual, the coal and oil industries are predicting that the cuts in greenhouse gas emissions agreed upon at Kyoto will cause severe economic hardship, and they are working hard to prevent ratification of the treaty in the United States Senate. Their arguments are reminiscent of the previously overstated claims that taking lead out of gasoline, replacing ozone-eating CFCs, and reducing sulfur emissions would ruin the economy.

Strong opposition in the United States Congress to the international agreement reached in Kyoto to reduce greenhouse gas emissions by only 7 percent is disheartening, to say the least. If this modest, insufficient step is met with such strong opposition, how can we hope to avoid the problems of global warming? The earth has already warmed and it will warm further before effective emission cuts are implemented. Just as a supertanker's momentum means that it cannot turn on a dime, the present momentum of our economy, which is based on the consumption of coal, oil, and gas, will not permit sufficient changes in ten years to prevent further warming from occurring. But just as the supertanker captain must anticipate a change in course well in advance, we must start to reverse the trend of ever increasing emissions of greenhouse gases. The Kyoto agreement is the first step in changing our course.

My optimism comes from the expectation that once we start moving in the right direction, technological progress will occur much faster than projected. As was the case for reducing lead, CFCs, and sulfur emissions, reducing greenhouse gas emissions may well be cheaper than the estimates used by industry naysayers to scare us. Once financial incentives are in place to conserve energy and to find other sources of energy that do not emit greenhouse gases, the entrepreneurial forces of technological innovation will surprise us by how quickly change occurs. As we have seen, several major environmental problems have been largely solved once we got over the inertia associated with the "we can't afford to clean up" mentality.

At the moment, this technological innovation is partially stifled by the low price of oil. Only a few companies have been motivated to design, develop, and market a fuel-efficient car while gasoline is so cheap. Even with this stifling of creativity, some new technologies using batteries and hydrogen fuel cells are advancing, albeit more slowly than needed. A similar situation exists for most industries. Why should industry change their manufacturing processes to reduce energy consumption and greenhouse gas emissions when the cost of oil, coal, gas, and electricity are so cheap? If they must buy greenhouse gas emission permits, on the other hand, as companies now must purchase sulfur gas emission permits, then the incentives for modifying old technologies or developing new ones will be stimulated. As with most new technologies, the cost will decrease as the demand grows and as new, cheaper innovations are developed. This scenario should apply to transportation, heating systems, electrical generation, and industry.

Is this optimism about technological solutions to global warming an example of "Custer's folly" described in Chapter 1? I don't think so, because new initiatives like tradable pollution permits are not equivalent to waiting for the technological cavalry to come over the hill in the nick of time. Advances in technology will be key to solving the global warming problem, but we cannot afford to wait passively for them to develop. Technology and entrepreneurialism need a boost to help us avoid further global warming in a timely manner. Once we shift our course, tacking toward an economy less dependent on coal and oil, the momentum of change could be dramatic, and hence my optimism. I have a hunch, for example, that the internal combustion engine that so quickly revolutionized transportation in the early twentieth century could become obsolete equally rapidly in the early twenty-first century, if only we take the right steps now to encourage development of the next generation of engines.

When we run out of oil that can be cheaply extracted sometime in the next thirty to fifty years, the switch to new technologies that do not depend on oil will undoubtedly occur. We may not run out of oil completely, but the sources that will remain in the year 2050 will not be the conventional oil fields that are easy to pump but rather will be tar sands and other types of deposits that are currently too expensive to extract. As conventional supplies of oil dwindle, the cost of oil will go up, and the incentives for innovation will develop because of the laws of supply and demand. Instead of waiting for cheap oil to run out, however, we need to jump-start technological innovation now. The same market forces will be at play in either transition scenario, whether driven by short supply of cheap oil a few decades from now or driven by government imposed, commercially traded greenhouse gas pollution permits starting any day now. If we value blue sky and the hospitable climate that it provides, the second choice is obviously superior.

A Gradual, Deliberate Scientific Consensus

A minority of scientists, mostly with the support of the coal and oil industry, espouse a dissenting view about global warming. It is important to understand why there can be room for debate when the vast majority of scientists agree that global warming is for real and very dangerous and why this majority view has been gaining ground only gradually over the past decade.

A consensus statement by scientists with such forceful clarity as the IPCC assessment was slow in coming because scientists are generally a conservative lot. The proof of an effect has to be very convincing before it is widely accepted by other scientists. Before concluding that global warming is caused by burning of oil, gas, and coal, all of the other possible explanations had to be reasonably ruled out first. It was determined that the earth had warmed, on average, about one degree Fahrenheit in the past century, but there were a number of plausible explanations for this warming. For example, it could have been simply that the measuring stations were located in areas that were distant from cities one hundred years ago but had since been swallowed by expanding cities. The "heat island" effect caused by the expanse of urban concrete and pavement could have resulted in warmer readings at the measuring stations. After extensive measurements and painstaking statistical analyses of the data, this possibility has now been conclusively ruled out as an important factor in the calculation of global temperature.

Another possible explanation was that natural variation in the strength of the sun's energy reaching the earth may have caused the observed warming. To most of us, the sun always seems like a constant, reliable source of warmth, but the amount of energy it produces actually varies somewhat from year to year, decade to decade, and across centuries and millennia. It took a long time for scientists to distinguish between this natural variation in the sun's warmth and the changes in warming caused by the greenhouse effect resulting from human pollution of the air. Variability in the sun's output probably has masked some of the effect of human-caused warming during the past few decades, but the human effect is now strong enough to detect it over and above the natural variation of the sun.

One of the reasons that some scientists were slow to reach consensus about human-caused global warming was that until very recently, the computer models for predicting climate change had some very troublesome flaws. Based on our understanding of the basic physics of how gases such as carbon dioxide trap heat within the atmosphere, the models predicted that the earth should have warmed about two degrees Fahrenheit during the past century instead of the observed warming of only about one degree Fahrenheit. Clearly, something was wrong with the computer models and our understanding of the warming effect, which left many scientists uneasy about proclaiming that they understood global warming well enough to make good predictions about the future or to lay blame on human pollution of the atmosphere.

In the past eight years, tremendous progress has been made in our understanding of why the earth has not warmed as much as expected given the huge changes we have made in the atmosphere. It turns out that other forms of air pollution are partially canceling the warming effect caused by greenhouse gases emitted when burning coal and oil. First, burning coal also emits tiny particles into the air that are called sulfate aerosols (the same thing that causes acid rain). In areas downwind of industrialized regions, like eastern North America and Europe, these particles are so abundant in the air that they reflect back into space some of the sun's energy, and so they have caused partial cooling of the Northern Hemisphere.

Huge plumes of dust blown from West and North Africa to the Atlantic, the Caribbean, and South America are additional sources of atmospheric particles that have a similar cooling effect. The savannas and deserts of Africa have been sources of dust for a long time, but the amount of wind-borne soil particles has increased in the past few decades as a result of increasing populations of people and their cattle and goats placing ever greater stress on overgrazed margins of the desert. As the plants that help hold down the soil are eaten, the exposed topsoil is blown away by the wind. This wind erosion in Africa is similar to the great dust bowl era of the 1930s in the midwestern United States, when a combination of farming marginal lands and drought caused tremendous dust storms, soil erosion, and loss of livelihoods. The effects of extensive soil erosion in Africa extend beyond the local impoverishment of soils to changes in the global climate.

Now that the importance of sulfate aerosols and particles of dust and soot has been recognized, they have been incorporated into the global climate models, and the model predictions of expected warming agree well with the measured global temperatures. The particles have had a cooling effect that partially offsets the warming effect of the greenhouse gases. This is both good news and bad. It is fortuitous that one form of pollution has been partially canceling out another form of pollution, but the outlook for the future is not so rosy. Emissions of sulfate aerosols are no longer rapidly increasing, because air pollution control efforts have been implemented to reduce acid rain, respiratory diseases, and other health problems caused by this form of air pollution, but the greenhouse gases are continuing to accumulate in the atmosphere. Unlike the past experience, the full warming potential of the still increasing greenhouse gases will be felt in the near future.

Sorting out this complex interaction of different types of pollution has taken time, and there is more to be learned. Studying the effects of different sizes and chemical types of aerosols is a very active area of current research, and the climate models will be further refined and improved as these research results are obtained. Our scientific understanding is never complete, but it is now widely recognized to be sufficient to make reasonable predictions about the magnitude of the warming effect in the twenty-first century.

Science for the sake of science must progress slowly and conservatively so that new ideas are fully tested before the old are discarded. For example, it may not matter from a practical point of view whether new theories in physics succeed in merging Einstein's theory of relativity with quantum mechanics to form a unified theory of physics within the next ten years or if it takes one hundred years. Although unforeseen technological advancements might result from such an advance in theoretical physics, nothing is now in danger that a new major development in physical theory would clearly save—better that the physicists take their time to fully test their new theories. But when our well-being and the earth's climate is at stake, we must act on the best scientific opinion that is available, even if it can only be expressed as "most likely." We cannot predict the future economic climate with complete certainty either, but we all make decisions every day about purchases, investments, and savings, based on whatever imperfect information we have at hand about the probability of future interest rate hikes, inflation, and unemployment trends. Similarly, we must decide now based on the strong probability of a warmer world that it would be prudent to change our fuel consumption habits. If we put off recognizing the importance of global warming and fail to change our use of oil and coal, we will commit the earth and our children and grandchildren to a warmer climate that will be hard to reverse. Whether we act now or fail to act now—either way—there will be consequences for decades to come.

LONG-TERM PRECAUTION VERSUS
SHORT-TERM PROFIT MOTIVE

Most of the dissenting opinions about the seriousness of global warming are driven by short-term profit motives. Ross Gelbspan has written a book *(The Heat Is On)* on the coordinated campaign by the coal and oil industry to cast doubt on global warming, and I will recount only a part of that story here. The most vocal of those handful of scientists still expressing a minority view about global warming have excellent funding from a group of coal and oil companies that goes by the misleading name the Global Climate Coalition.

For those who want excuses not to heed the call for action supported by the majority scientific view, the Global Climate Coalition has provided plenty of misinformation to fuel delays. Like the tobacco companies fighting restrictions on cigarette use, some of the oil and coal companies could have a lot to lose if global warming is taken seriously by the public and by policymakers in government. This small group of industry-supported scientists is also very clever at getting their editorials published in newspapers, their books marketed forcefully, and their views espoused by radio talk show hosts.

The media likes to present both sides of any issue as if they were boxers of equal stature and strength, and so scientists with opposing points of view are interviewed as if they held equal stature and respect within the scientific community. In terms of strength of argument and credibility, the IPCC scientific consensus about the importance of global warming is a heavyweight compared to the bantam weight of the handful of dissenting scientists. Unfortunately, the well-funded and ideologically and financially motivated bantams are running circles around the pensive, cautious, lumbering heavyweight, and the impact of the bantams' clever program of misinformation far exceeds their numbers or their scientific credentials. Their strategy has been to find little chinks in the armor of the global warming evidence, draw attention to these minor points, blow them out of proportion, and thereby gain publicity in the popular press that casts doubt on the strong mainstream scientific consensus on global warming. When subsequently debated in the peer-reviewed scientific literature, these issues are usually put to rest, but by then, the damage has already been done in the popular press, and the global warming naysayers achieve their goals of undermining confidence in the science behind the global warming consensus.

The aggressiveness of the Global Climate Coalition may be backfiring, however, as several of the corporate members have withdrawn their membership in protest. "Global climate change is a serious problem and we need to take steps to deal with it," said an executive of the Arizona Public Service Company. An executive of BP America, a subsidiary of British Petroleum, explained BP's withdrawal by saying that it will instead work through "a more moderate and conciliatory body." Shell International, BP/Amoco, and Sonoco have made pledges to reduce their greenhouse gas emissions. These companies have joined several others to form the Business Environmental Leadership Council, organized by the Pew Center on Global Climate Change. Rather than denying the reality of global warming, as the majority of the petroleum industry continues to do, this responsible business group has issued a clear statement accepting the compelling scientific evidence: "We accept the views of most scientists that enough is known about the science and environmental impacts of climate change for us to take actions to address its consequences." These corporate leaders have endorsed the Kyoto agreement and urge businesses to help "meet emission reduction objectives, and invest in new, more efficient products, practices, and technology" (for the full statement, see their Web site: http://www.pewclimate.org).

Despite these examples of corporate responsibility, other corporations are still joining the old energy lobby coalition, which continues to orchestrate a campaign of misinformation and denial about the importance of global warming. The analogy with cigarettes is worth repeating. Over thirty years ago, the surgeon general of the United States, with the support of a strong consensus among doctors and medical researchers, warned that smoking is hazardous to one's health. At first, the tobacco industry brought out their own scientists to present the minority view that smoking is not related to illness in an attempt to refute the surgeon general and to cast doubt in the minds of smokers. Truth eventually overwhelmed this initial strategy to confuse the issue in public; yet until 1997, the tobacco companies refused to admit that cigarette smoking causes lung cancer and heart disease.

We are also addicted to our cars and to our use of oil, gas, and coal. The scientific community, like the surgeon general, has clearly warned us that our addiction is harmful to our health and to the earth's health. The coal and oil industry denies this warning and is sending out its own scientists to debate and confuse the issue with misinformation. Perhaps these people really believe what they are saying, because they want to believe it, and the human power of rationalization and denial of the truth is amazingly strong. Truth will eventually prevail in this case too. By then, however, the earth will have already warmed significantly more, creating serious problems that could have been averted if reason had ruled early instead of greed.

The misinformation job of the energy industry apologists is made easier by winter storms. My neighbor once teased me after a big winter storm: "You still believe in that global warming stuff?" Paradoxically, big snowstorms are entirely consistent with global warming. The expected changes in climate will make winter both milder in terms of average temperatures and occasionally more harsh in terms of blizzards and freak storms with lots of snow and freezing rain. Remember that it seldom snows when it is very, very cold. Big snowstorms need moderate winter temperatures and lots of energy to mix moist air masses originating from the ocean with cold air masses from the Arctic. The blizzard of 1996, which dumped two feet of snow on Philadelphia and much of the east coast of the United States, occurred because cold Canadian air had been brought unusually far south by the jet stream and it mixed with warm moist air from over the Atlantic to produce record-breaking snowfall. This mixing of cold air and wet air to produce snow requires energy, which global warming will provide with increasing frequency. We cannot say that this particular storm was necessarily caused by global warming—maybe it was, maybe it wasn't—but it is a good example of the unusual weather that is likely to come. So don't expect that global warming will mean you can put away the snow shovel.

Nor should anyone expect cold spells to go away completely. They may become less common and predictable,

> The more rapidly we force changes in the [climate] system,
> the more likely it is to exhibit inscrutable behavior.
> —**Stephen H. Schneider, Stanford University**

but they will continue to occur because the world's weather is by no means uniform. On average, the earth can be warmer while there are still intensely cold spots in localized regions for periods of time. We can expect the weather to be the opposite of the Boston Red Sox team record. The Sox are sometimes hot and sometimes cold. Their winning streaks come often enough to tease the Boston fans, but in the average year they usually fall into a cold spell just before the season is over. Conversely, the weather will be sometimes hot and sometimes cold, and the cold spells will still come often enough to tease us with some real winter, but on average, winters will be increasingly mild and summers increasingly hot.

Curiously, the coal and oil industry softens its objections to reducing greenhouse gas emissions when the discussion turns to requiring the auto industry to improve gas mileage on new cars. Of course, the auto industry does not remain in neutral on this issue, as they have strongly resisted all efforts to regulate fuel efficiency. In fact, after some progress in the 1970s, the average fuel efficiency of the American auto fleet has actually declined due to a shift in preference to trucks and sport utility vehicles. The technology exists to double fuel efficiency of all vehicles or perhaps improve it beyond that, but the auto industry has successfully fought stronger government regulation of auto fuel efficiency for fear that it would hurt profits. The argument is reminis-

cent of the fear that taking the lead out of gasoline would
hurt profits (and the GNP). The auto industry argues that
if global warming is a problem, we should tax coal and oil
rather than burden the auto industry. As the buck gets
passed back and forth, the cost of global warming will
eventually be borne by us all.

Just as a few enlightened petroleum companies have recently started to take some important initiatives to reduce
their greenhouse gas emissions, a few automobile manufacturers have also started to act more responsibly. Ford
Motor Company is one of the growing number of corporations that has pulled out of the Global Climate Coalition
(the industry group that continues to deny the reality of
global warming). Toyota has joined the Business Environmental Leadership Council of the Pew Center on Global
Climate Change (the business coalition that endorses the
Kyoto agreement to reduce greenhouse gas emissions).

More importantly, these pledges and position statements
are being backed up with meaningful actions by a few auto
manufacturers who are addressing the vast gulf between
the unimpressive gas mileage of the majority of vehicles
currently on the road and the much better mileage that is
technologically possible. Already in use in Japan are
Honda and Toyota sedans that give double or triple the
common gas mileage by employing hybrid engines. By
switching between power from batteries and power from
combustion of diesel fuel, or using both, depending on the
operating conditions, the hybrid engine is able to run most
of the time at low revolutions per minute, which is near
peak efficiency. Some of the energy normally lost during
braking is harnessed to recharge the batteries. A few thousand of these sedans are being marketed in the United

States and Europe starting in the year 2000, initially at a price below cost, in an attempt to stimulate interest in this new market while further technological developments bring down the manufacturing costs. Ford and Daimler-Chrysler are working on hybrid engines for sport utility vehicles (SUVs) that would make modest improvements on the miserable mileage SUVs currently deliver; these vehicles should be available by the year 2003. Daimler-Chrysler is lobbying the U.S. government to create a tax deduction for purchases of these vehicles, which would help make them more financially appealing to consumers.

In this case, part of the private sector is moving in the right direction, although encouragement from governments and interested consumers is needed to spur on and accelerate these developments. In addition to regulating fuel efficiency, or perhaps in lieu of regulations, financial incentives for developing and buying fuel-efficient engines would also help achieve the goal of reduced greenhouse gas emissions.

BACK TO EVERYDAY ECONOMICS

The skeptic who remains unconvinced about global warming would do well to take a clue from an independent, profit-motivated, conservative market player. The insurance industry is not interested in global warming for any ideological reasons. It is out to make money, and it is expert at assessing risks in such a way as to keep making money. The insurance industry was not fooled by the tobacco industry's claims that the danger of smoking was unproven. For many years, smokers have been charged a lot more for insurance because smokers are at much greater risk of dying young, getting sick, burning down their houses, and even getting in automobile accidents. Now the insurance industry has realized that owning property next to the coastline is increasingly risky because global warming means that the oceans are rising and that storms may be increasing in number and intensity. The insurance industry will probably also lead the corporate world as they factor in expected changes in climate when calculating premiums for insuring farms, dams, sawmills, or any structure or business that is immediately susceptible to changing climate.

In fact, Munich Re, one of the world's largest reinsurance companies (the companies that insure the insurance companies that sell us insurance), has called for political action to control global warming, which it blames for recent increases in claims related to natural disasters. An uncertain effect of climate change is that it may cause increased incidence of hurricanes, like the 1995 hurricane season, when only the names Van and Wendy remained unused at the end of the alphabet. Likewise, the El Niño event of 1997 and 1998 drenched Peru, western Mexico, and California with torrential floods and scorched Indonesia, Mexico, and Brazil with drought that touched off devastating forest fires. This El Niño may have occurred without global warming, but global warming may have contributed to its severity. In fact, the evidence is accumulating that global warming will contribute to increased frequency and severity of future El Niño events.

A spokesman for Munich Re recently left little doubt about how it is calculating the economic risks of global warming: "We cannot afford to wait for definite scientific proof of global warming, and would be better advised to be prepared for the worst." Ski resorts have also seen the future, and they are investing in snowmaking machines because snowfall has already become less dependable and interspersed with winter warm spells.

I find it sad to close this chapter by citing insurance underwriters and ski resort operators rather than scientists as the most objective and authoritative spokespersons for the genuineness of the threat of global warming. Scientists have not done a good job of speaking directly to the public, and the oil and coal industry's hired scientific guns have succeeded in making the scientific community look uncertain. But the science is there and it is strong. Perhaps most nonscientists can relate better to the idea of risk when it is expressed in everyday economic terms of insurance premiums than when it appears as a range of probabilities of warming rates expressed in the jargon of scientists. In any case, our climate is at serious risk. To return to the medical analogy of high blood pressure, the first step toward dealing with a diagnosis is to acknowledge that the problem exists. Unfortunately, much of our society seems to be stuck in the denial phase and has not accepted the diagnosis. There are many ways, including the new tools of ecological economics, to place monetary and nonmonetary values on blue sky in order to avert the risk of global warming. But first, we must get past the denial phase and accept that the problem is real, imminent, and serious. Most importantly, we must recognize that the air is an invaluable and irreplaceable resource that is essential for our ecological and economic well-being.

6 🌿

Global Garbage

Malthus Revisited

IN DISCUSSING SOILS in Chapter 2, I explained how Thomas Robert Malthus's nineteenth-century prediction of massive famine resulting from exploding human population was averted, or at least delayed, through the inventions of modern agricultural technology, such as chemical fertilizers, pesticides, and the breeding of new varieties of crops. And it is true that these technological developments have worked wonders. But they have also created huge problems that are every bit as big as, or even bigger than, the problem of food scarcity. To explain this phenomenon, I propose two new laws.

THE LAWS OF TECHNODYNAMICS

1. Conservation of problems: Problems do not go away, they are merely substituted, one for another. The solution of one problem creates another problem.
2. Technological challenges always increase. As the human population increases and natural resources

remain constant or degrade, then technological challenges will increase in size, number, and complexity.

Those who can remember anything from high school chemistry might recognize my facetious analogy to the first and second laws of thermodynamics (conservation of mass and increases in entropy). Making analogies between human social behavior and laws of nature is questionable, but these laws of technodynamics are convenient ways of illustrating a point about the interplay between technological development and management of the environment.

There is little doubt that technology can solve many of our environmental problems, but it is often forgotten that each technological solution usually creates new unanticipated problems. For example, horse-drawn buggies were still the most common form of transportation in 1910, and New York City was having problems disposing of the huge volumes of horse manure being collected on its streets. Some thought that the accumulating manure would reach crisis proportions and could see no simple solution. But the advent of the automobile made the manure issue moot, as horse-drawn buggies became obsolete and the horses were put to pasture. This story has been used in an attempt to paint today's environmentalists as alarmists, similar to the New York inhabitants who were once worried about too much horse manure. This line of logic argues that just as technology in the form of newly invented automobiles solved the horse manure problem, technology will also solve all of our current environmental problems, and so we should not worry about the environment.

The flaw in this logic is that today's environmental problems are larger in scope and more severe than the localized New York City horse manure problem of 1910. Although the technological development of the internal combustion engine certainly solved the local manure problem in New York, we must not forget that automobiles have since created a giant global problem of too much carbon dioxide and nitrogen oxide garbage in the atmosphere coming from all of those car engines burning gasoline. Leakage of oil and fuel from underground storage tanks have contaminated groundwater. Another future technology will eventually replace internal combustion engines and other uses of oil, but until then, we have a global problem that is already affecting the climate and that is going to be more difficult to solve than was the rather well-defined and localized problem of manure in New York City. One problem has been substituted for another, and the new one is bigger, global in scope, and more difficult to solve. We can solve it, but not by complacently waiting for the technological cavalry to come over the hill in the nick of time. The detrimental effects of these newer, bigger problems on both the environment and the economy will be less severe if we can get a head start on developing the technologies to solve them.

GARBAGE AS AN EXTERNALITY

I am using the term garbage in a broad sense, to include all of the waste products of agriculture, industry, and consumerism that we throw away or allow to seep, diffuse, or otherwise seemingly disappear into the environment. The New York City horse manure story is an easy one to follow regarding one form of garbage produced by the type of transportation commonly used in 1910. The garbage problem that this chapter addresses for the late twentieth and early twenty-first centuries is the regional and global pollution of water, and ultimately of groundwater, resulting from our modern technological solutions for feeding the world's enormous human population.

Without modern chemical fertilizers and pesticides, there would have been much more suffering and starvation during the past century; that is a fact. Unfortunately, it is very difficult, perhaps impossible, to apply only enough fertilizer and pesticide to nourish and protect only the crop plants. The farmer routinely applies more fertilizer than the plants need because he knows that a large portion of the fertilizer, often more than half, will be washed away by the rain before the plants can use it. As long as technology provides cheap fertilizer, this inefficiency is acceptable to the farmer because the cost of the wasted fertilizer is more than made up for by the better yield of the well-fertilized crop. The farmer usually does not worry much about where the unused fertilizer and pesticide go, because once they are off his property they are no longer his problem. The neoclassical economist usually considers the cost of water pollution to society as being *external* to his calculations of agricultural economics, just as it is off-site for the farmer. Some economists study these *externalities;* others tend to ignore them.

Some of the fertilizer washes into the streams and lakes where it becomes the problem of the people who live downstream and anyone who cares about the health of the rivers and lakes. The fertilizer that makes crops grow better while still on the farmer's field also makes unwanted algae grow better in streams and lakes, which chokes out the native aquatic plants, depletes oxygen in the water that the fish need, and destroys the natural food chain. Some of the fertilizer and pesticide also make their way into the groundwater, which is out of sight but a vital source of drinking water and irrigation water in many regions of the world.

Notice that the farmer's first problem—how to make his crop grow better—has been solved by applying modern technology's chemical fertilizers and pesticides, but a new problem—pollution of streams, rivers, lakes, and groundwater—has been created. Moreover, the problem is no longer a local one confined to a farmer's field; it has grown to a regional one that affects everyone who lives downstream or who uses groundwater. Local problems are almost always easier to solve than are regional ones.

An example of solving local water pollution problems occurred in the United States in the 1970s, when we made great progress in reducing pollution from localized point sources, such as the outflow of sewage plants and factories. As a result, the Cuyahoga River in Cleveland, once one of the most polluted rivers in the United States, has not caught on fire in years, and people can now swim in Lake Erie. Gregg Easterbrook cites this progress in his recent book to argue that the environment is no longer in serious danger. Easterbrook's message is an unfortunate example of allowing a little bit of progress to give us a false sense of security, prompting us to fall into the third fallacy described in Chapter 1: false complacency, or not beating the wife, or the earth, as much as before.

In this case, Easterbrook's analysis is faulty because he apparently does not understand the depth (both figuratively and literally) of the water pollution problem. Although it is true that many rivers and lakes can and do recover within a couple of decades if the major point sources of pollution can be identified and successfully cleaned up, the diffuse *nonpoint* pollution sources coming from hundreds of farmers' fields have proven to be much more difficult to control. There is hope of eliminating even these pollution sources to rivers and streams, because we now know that creating strips of forests along riverbanks will help absorb and transform the pollutants coming off of the farmers' fields before these pollutants enter the streams. According to the second law of technodynamics, however, this solution will be more difficult than the solution to the first problem (getting the crop to grow better), because the owners of land along the riverbanks must be convinced that they should take land out of its current use and devote it to strips of forested land for the sake of filtering out pollutants, and there is presently no monetary incentive to do that. Perhaps ecological economics can devise a clever way of creating a market for filtration of pollutants entering streams and rivers.

GROUNDWATER POLLUTION AND DEPLETION: THE INSIDIOUS EXTERNALITY

Easterbrook also fails to see the more important problem of groundwater pollution and depletion. The groundwater problem is yet another order of magnitude more difficult to solve than is pollution of rivers and lakes. Just as we can see local smog air pollution but not the greenhouse gas air pollution that is causing global warming, we can see dirty rivers and lakes, but we have few reminders in everyday

As water is so vital to every ecosystem and every human endeavor, one might expect that modern societies would be mindful of the limited fresh water resources available to them, and take the utmost care to ensure their continuing availability and high quality. Unfortunately, that is seldom the case. The cost of drawing water from a source seems so low that few people are aware of the potential cost of replacing the water after the source has been exhausted, or of purifying it after it has been fouled. Much as they continue to degrade their soil resources, modern societies go on depleting and contaminating their water resources, in mindless disregard of the consequences.

—Daniel Hillel, University of Massachusetts

life about the more insidious, pervasive, and dangerous problem of groundwater pollution and depletion.

Enough water flows through rivers and lakes to flush them out within a few years or decades, so that if a pollution source is stopped, nature's flushing action will help clean up the river and lake water, but this is not true for most groundwater. First, the amount of groundwater that is threatened by pollution is astounding. The water held underground is about one hundred times greater than the water in all of the lakes in the world combined. So as this huge reservoir of underground water becomes polluted, we will have a much bigger cleanup job than we had for rivers and lakes. Second, this huge reservoir of underground water is flushed out over several hundred or thousand years or more (the time varies depending on which *aquifer*—underground water storage area—we are talking about). So if we find that our groundwater has become polluted, we cannot simply stop the source of the pollution and then expect the problem to go away within a few decades, as it has partially done for the Cuyahoga River and Lake Erie. Instead, it will take hundreds or thousands of years for nature's flushing action to clean up the groundwater.

As in the case of soil erosion, there are no comprehensive global databases available to calculate with much confidence the rate at which groundwater is becoming depleted or polluted. Just about everywhere that groundwater has been studied, however, problems have been identified, and the experts in groundwater studies are seriously worried. In a workshop of economists, sociologists, geochemists, biologists, and hydrologists convened in 1997 by the National Research Council to study the "transition to sustainability," my colleagues and I came to the conclusion that wherever rainfall is high enough to recharge the groundwater, there are usually concerns about the groundwater becoming contaminated with agricultural or industrial chemicals or human sewage, and wherever the climate is arid, local communities and farmers are usually withdrawing water from the ground faster than it is being replenished by rainfall. There may be some remote areas without significant human activity that have neither problem, but wherever humans are concentrated, the groundwater appears seriously threatened by one or both of these problems.

Plumes of pollutants have been identified in groundwater throughout the United States and all over the world. In a recent analysis of well water in the United States, 25 percent of the wells in one-third of the counties had elevated nitrate concentrations. In 5 percent of the counties, the measured values for nitrate in the contaminated wells were above federal drinking water standards, thus posing a direct threat to human health. Still more troublesome are the complex chemical residues of pesticides and fuels from leaky underground storage tanks that decompose very slowly. Although some of the most dangerous pesticides have been prohibited in the United States, many are still exported and their use is increasing in other parts of the world.

Elaborate and expensive schemes have been devised to pump the contaminated groundwater out of the ground and through some sort of treatment facility. For some of the smaller plumes of contamination by chemicals that can be readily removed by these methods, this approach works. In many of the larger and more difficult cases, however, the cleanup job requires new research to develop and to test new technologies. For example, an aquifer contaminated by jet fuel on a military base near my home on Cape Cod involves several plumes of different chemicals, some of which we know how to clean up, but some of which require field trials of untested technologies. For one of the plumes, there had been a plan to use the conventional pump-and-treat method, until it was realized that more water would need to be pumped out of the ground for the cleanup than is currently being pumped out of all of the wells in all of the communities on Cape Cod combined. This massive pumping would have altered water levels of ponds and wells and would have changed the flow patterns of the underground aquifer. Once again, the solution to the first problem—a contaminated aquifer—would cause another problem—locally depleted groundwater from too much pumping.

The current plan of action for the polluted Cape Cod aquifer is to try a couple of new experimental systems: One will force air through the groundwater within the well, which will hopefully allow the volatile chemicals to escape into the air; and another system will use a "wall" of iron filings in the ground, which will hopefully break down the toxic chemicals as the groundwater naturally flows past it, thereby avoiding the need to pump out the water. As a resident whose drinking water depends on this aquifer, I sincerely hope that these new technologies work, but the real take-home message of this lesson is that, frankly, no one really knows how to clean up this groundwater pollution problem. Our problems on Cape Cod are not unique. The Department of Defense has a budget of more than $1 billion to work on similar cleanup projects across the country.

A frightening situation has recently emerged in India, near the border with Bangladesh. More than one million people are dependent on well water there, but a recent large survey of the well water revealed that most samples were contaminated with unacceptably high levels of arsenic. The study was prompted by massive outbreaks of skin lesions and skin cancers that have been directly linked to arsenic poisoning. An estimated 200,000 people already have skin lesions.

Arsenic occurs naturally in the deep soils and sediments of this area, but it normally stays in a form that is not soluble and does not enter the water. Why it is entering the well water now is not entirely clear, although there are two leading hypotheses that are both related to human activity. One possibility is that phosphorus fertilizers from expanding and intensifying agriculture in the area are leaching into the groundwater, where the phosphorus releases the arsenic that would otherwise have stayed bound to the deep soil minerals. Another hypothesis is that the growing population and the intensified agriculture needed to support that population have withdrawn so much well water from the ground that more oxygen can now penetrate deep into the ground, where it transforms the arsenic into a soluble form. Changing agricultural practices so that less fertilizer and less irrigation water are used might help stop the release of arsenic into the groundwater, but we have no idea how long it would take for the arsenic concentrations to come down to acceptable levels, and the people must grow crops and eat and drink in the meantime. The only short-term solution is to build pipelines to bring in water from the Ganges River, which is estimated to cost $200 million, not to mention the nonmonetary costs of diverting water from the Ganges ecosystem. Once again, the local problem becomes a regional one with increasing complexity.

GROUNDWATER DEPLETION:
YOU CAN'T DRINK GNP EITHER

Groundwater pollution is not the only worry; in many areas the more important concern is that it is being used up too fast. As we pump water out of the ground faster than it can be replenished by rainfall, the amount of groundwater declines. Although the amount of water in the ground is astounding, the rate that we are pumping it is also amazing. About 200,000 wells in the Ogallala aquifer under the Great Plains states from Texas to South Dakota have been pumping so much water that at one point the water table was dropping by four to six feet per year. In the meantime, rainfall replenished the groundwater by only about two inches per year in the southwestern portion of the region. Not surprisingly, the cost of pumping this water has increased as the water table has dropped, and pumping for agricultural irrigation has declined by about one-third since 1974. Similar examples can be found throughout the world. In an area of northern China inhabited by about 100 million people, the water table has fallen thirty meters during the past three decades.

In the case of the Ogallala aquifer under the Great Plains of the United States, marketplace economics related to the cost of digging deep wells provided some impetus toward a solution to the problem, albeit well after the problem had become acute. A full solution has not yet been realized. Because the groundwater was and still is underpriced in the neoclassical economic system relative to its crucial role in the ecological system, groundwater depletion will continue until it is extremely scarce or until the current market system is modified. Moreover, the ability of future generations to utilize groundwater for the irrigation that will be needed to adapt to the warmer and drier climate predicted for this region as a result of global warming has already been severely diminished by previous mismanagement and depletion of the groundwater resource. Unfortunately, this type of overextraction of groundwater supplies is common throughout the world.

You don't have to be a rocket scientist to understand that pumping water out of the ground faster than it is being replenished by rainfall will eventually lead to empty wells. My son figured out at age two that opening the bathtub drain meant that his bath would soon be over, so why is it that politicians and government officials do not understand that pumping out the groundwater aquifers will leave us as helpless as a dry rubber ducky? How can we be so short-sighted?

Tragedy of the Commons

Although some states legally recognize the right of landowners to pump and use groundwater, which is tantamount to a form of ownership, groundwater does not obey property boundaries. Aquifers are usually large and they span across many properties, counties, and even states and nations. Hence, the groundwater resource is a *commons*—a shared resource. But is everyone taking only his or her fair share? When I sink a well on my land I may be sucking up water from underneath my neighbor's land too. If my corporation sucks up water faster than anyone else to irrigate my giant corporate farm, then I will get more water than anyone else and my operation will be more profitable. Governments may choose to regulate how much water is withdrawn from the ground if they so desire, but they generally have not done so, at least not effectively. The reasons include inept governments, angry landowners (read voters and campaign contributors) who do not want to be told by the government how to operate wells on their land, and greedy corporations that buy influence at all levels of government. Most engineers, economists, and ecologists understand this problem, but they are ignored. Those who have used the engineers' relatively simple calculations to speak out that this foolishness and greed cannot go on forever have been effectively labeled as environmental extremists. As long as water flows from our taps, most of us remain unconcerned. Modern plumbing keeps us comfortably insulated, for the moment, from the dwindling resource upon which our long-term comfort and well-being depend.

When Garrett Hardin first presented his notion of the *Tragedy of the Commons,* he used the example of community-owned and shared pastureland, such as the town commons that could be found in many eighteenth-century New England towns. The local residents could graze their cows there, and each person was expected to use this resource responsibly. In this system, however, each farmer has an incentive to cheat on his neighbors by letting his cows graze a bit more than their fair share of the common pasture. If everyone cheats this way, then the pasture will be overgrazed and become degraded. Groundwater is also a commons that is shared by an entire region, and each user has an incentive to cheat by overpumping it or by allowing his agricultural, industrial, or residential garbage to seep into it.

That the tragedy of the commons is inevitable because of our inability to resist cheating on our neighbors has been challenged by supporters of community fisheries. Fish in rivers, lakes, estuaries, and oceans are another type of commons. The loss of fish populations has been tragically extensive, and the causes of this tragedy of the fisheries commons is being widely debated (see Chapter 9). Those who study communities that rely upon fisheries for their livelihoods note that people do talk to each other in communities and they do influence each other's actions in a variety of ways. From local peer pressure to governmental regulations, society can use many tools to make sure that the commons are well managed and that cheating is not tolerated. Peer pressure and enforcement of local regulations may have been sufficient for some of the eigh-

> An ethic, ecologically, is a limitation on freedom of action in the struggle for existence. An ethic, philosophically, is a differentiation of social from anti-social conduct. These are two definitions of one thing. The thing has its origin in the tendency of interdependent individuals or groups to evolve modes of co-operation. The ecologist calls these symbioses. Politics and economics are advanced symbioses in which the original free-for-all competition has been replaced, in part, by co-operative mechanisms with an ethical content.
>
> **—Aldo Leopold**

teenth-century town commons, but proper management of fisheries and of the threatened, essential groundwater commons of the twenty-first century will require both regulatory and economic incentives for users and polluters.

Who Pays the Global Garbage Man?

In most communities in the United States, local town and county governments are responsible for maintaining the local garbage dump and for paying the salaries of the workers who pick up garbage at the curbside. This is a widely accepted role for local government. What about regional and global garbage? Who is responsible for managing the disposal of garbage into the groundwater and into the atmosphere? Who will pay the "global garbage man"?

When an environmental problem is confined to a locality, as in a county landfill site or a toxic waste dump on property owned by a specific individual, then the responsible party is usually clearly identified. But if that individual can get away with letting those toxic materials "disappear" into the rivers, groundwater, or the air, then society as a whole, rather than the individual, must deal with it. Of course, the first laws of thermodynamics and techno-dynamics assure us that those toxic substances do not actually disappear—they are merely transferred to a new place and to a new owner of the responsibility. When the locally confined problem is dispersed to a larger area and where responsibility is less clear, the problem is nearly always more difficult to address. The biggest and most difficult problems are those that have been transferred from the responsibility of individuals to the responsibility of society in general.

If the resources at risk of depletion and pollution are not owned by anyone—the Big Sky of Montana, the Ogallala aquifer of Texas and much of the rain forest of the Amazon are not owned *per se*—then government must play a significant role in their management. Conservative and liberal ideologies may espouse different views of *how* government can best achieve these goals. In some cases, the role may be governmental regulation. In other cases, private enterprise can be prodded to help protect these resources by using the clever techniques of ecological economics discussed in previous chapters, but Adam Smith's invisible hand will not do it alone without government encouragement or insistence.

It should also be recognized that bad government can create greater problems than no government intervention at all. Although the nuclear accident that we all know as the Chernobyl disaster is infamous, there is an equally tragic, foreseeable, unnecessary, and environmentally disastrous legacy of the government of the former Soviet Union. The central planners of the Soviet government thought that they could develop the arid plains of the south central Asian region by diverting the water that normally flows into the giant Aral Sea. This diverted water was used to irrigate a massive new cotton crop that would benefit the Soviet GNP as well as the people of the region.

Any soil scientist with a whit of experience would have predicted that the irrigated land would become salty as the irrigation water evaporated. Any engineer, hydrologist, or child in a bathtub could have easily foreseen that the Aral Sea, robbed of its incoming water, would shrink in a matter of years. Many of these lands are now so salty that they can no longer be tilled, agricultural chemicals have polluted the groundwater that is used for most drinking water in the region, and the Aral Sea, once the world's fourth largest lake, has lost about two thirds of its water, and its shoreline has receded some twenty-five miles. Once teeming with fish species found nowhere else on earth, the Aral Sea has now become a salty brine, and most of the native species of fish are gone. The human tragedy is similar in magnitude, as the new agricultural endeavor largely went bust while also destroying the traditional farming and fishing ways of life.

Soviet socialism is by no means the only form of government that has been guilty of intervening with disastrous policies. As discussed in Chapter 2, U.S. government subsidies of cheap water in the American West are largely to blame for selenium poisoning of wildlife and salinization of soils. Fertilizers and pesticides are subsidized by the governments of many developing countries, which results in their overuse and the attendant pollution. On the other hand, government can also play a positive role by creating ecological economic incentives for water conservation, for establishing forested buffer strips along stream banks, and for using pesticides and fertilizers more judiciously. In some cases, old fashioned regulations may be necessary to insure that groundwater pumping does not exceed recharge rates, thereby preventing the tragedy of the commons.

DESIGNED RECYCLING: AVOIDING THE GARBAGE PROBLEM IN THE FIRST PLACE

Most of us are familiar with the recycling of bottles and cans. In some states, we pay a nickel deposit for every bottle of soda or beer when we purchase it. The nickel is returned when we return the empty bottle for recycling. At a fair in Frankfurt, Germany, an acquaintance of mine paid an extra mark for a cup of coffee. When the empty cup was returned, she got her mark back, and the coffee cup was reused. Although these are excellent examples of designing reuse of products, by no means should we restrict the idea of recycling to reusable beverage containers.

A novel approach to the garbage problem has been offered by architect William McDonough, who sees design as the key. Instead of producing carpet, for example, that is designed to last five to ten years and then sent to a landfill when it is worn out and replaced, what if the carpet manufacturer designed a carpet that was meant to be recycled? Instead of buying a carpet, the homeowner would essentially lease the carpet, with the intent of "trading it in" for a new carpet when it needs to be replaced. The vendor would ship the worn out carpet back to the manufacturer, who would be expecting it and would already have a plan for recycling the fibers within the carpet. Under this scheme, the manufacturer is less likely to include toxic chemicals in the original product if he knows that he must recycle them when they are returned to him.

In contrast, the type of afterthought recycling that is somewhat common today requires the development a new recycling industry to use up the recycled waste of another industry. For example, new companies have been created to weave sweaters out of the plastic fibers recycled from old milk containers. Maybe this will catch on, but fibers from plastic milk jugs may not be the first material of choice to be worn on the human skin. The ability to find economically viable uses of recycled plastic milk containers might be improved if the milk company already had a recycling application in mind when it designed the type of plastic to be used in the milk container in the first place.

This new type of recycling would begin when the product is originally designed by the manufacturer. Carpet and milk containers are good examples, because we are all familiar with how the large and unwieldy old carpet must be hauled away to the garbage dump and with how many milk containers we throw away or recycle every month. Rather than trying to promote a new industry to find a use for carpets otherwise destined for the garbage dump, the carpet manufacturers, themselves, could design their carpets in such a way that they would be able to reuse them five to ten years later. Built-in recycling designs could apply to automobiles, computer components, and just about any consumer product that is known to have a predictably finite functional lifetime. This emphasis on design would not only save space in landfills but also reduce probabilities of leaching pollutants to rivers and to groundwater. If the manufacturer knew that the product was coming back to him, he would be less likely to put a toxic compound into it that would be difficult to deal with later. Any chemical, fiber, or compound that is designed to be reused is one less bit of garbage that will be released to the land, air, or water.

How can manufacturers of carpets, milk containers, computers, or any other product be encouraged to design the reuse of their products? As much as designed recycling makes sense from an ecological perspective, it will not happen until there are economic incentives. Ecological economics will need to find innovative ways for providing economic incentives to manufacturers. Taxes and tax credits are obvious tools, although the T-word is often politically unpopular. How much would you be willing to pay for a carpet that has a design for its eventual recycling built into its purchase price, compared to the cost and inconvenience of sending the worn out carpet to the garbage dump several years later? Economists would do us a great service by researching ways that designed product recycling could be encouraged through financial incentives.

MALTHUS IN THE TWENTY-FIRST CENTURY

What would Malthus have said about our current problem of accumulating garbage on land, in the atmosphere, and in the groundwater? Modern agricultural technology has forestalled Malthus's predicted famine for now, and maybe future technological developments will keep famine down to a dull roar in the future as the world's population climbs from its current 6 billion people to 8 or 10 billion or beyond (although the ability to feed that many is far from certain). But clearly, technology has merely transformed the food shortage problem into a global garbage abundance problem. Whether it is carpets in landfills, fertilizers in stream water, pesticide residues in well water, or greenhouse gases in the atmosphere, we must deal with the huge quantity of waste produced through feeding, clothing, and housing 6 billion people, with about 90 million people added every year.

The first step in solving the groundwater pollution problem is to recognize that we still depend on clean water, that the water resource is being mismanaged and is in severe danger, that technology cannot substitute entirely for nature's cleansing of the water resource, and that there is no time for complacency. The threat to groundwater is extremely serious because we depend on it so extensively, because it is so hard to clean up once it is contaminated, and because it takes thousands of years to replenish the groundwater once it is depleted. More than any other environmental problem that I can think of, an ounce of prevention of contamination and depletion of groundwater is worth gallons and centuries and megabucks of cure.

A new Malthusian doctrine is needed that recognizes the first and second laws of technodynamics and that places as much emphasis on dealing with the garbage created by the growing population as it does on feeding the increasing number of mouths. As the human population increases further, stress on the environment due to disposal of all kinds of industrial, agricultural, and human wastes will increase exponentially. Technological advances for dealing with this kind of global garbage will help—indeed, they will be essential—but the technological solutions for both feeding this population and disposing of its garbage will become increasingly more challenging. Unlike the old Malthusian doctrine that predicted massive famine, the new one sees food production as a challenge, but it also predicts very difficult waste disposal problems. Also unlike the old Malthusian prediction, disaster is not inevitable. The neo-Malthusian view of the future has plausible happy outcomes, but only if steps are taken now to avoid those forms of pervasive, long-lived pollution, such as groundwater contamination, that require centuries to reverse.

7 🌿

In Search of Sustainability

From Small Landholders to
Macroeconomists

BY HOLDING DOWN THE right foot pedal on a piano, you can release the keys after playing a chord and listen to the harmony resonate for a minute or two or more as the pedal sustains the chord. While sitting there on the piano bench listening to each pitch gradually fade, it can seem like the notes keep ringing indefinitely, albeit ever so faintly. The right foot pedal is called the *sustain* pedal.

A popular buzzword these days in the realms of agronomy, forestry, fisheries, and development is *sustainability*. One goal of sustainable agriculture, like the sustain pedal on the piano, is to try to keep agriculture going as long as possible. In other words, we should not do anything now, like let the soil erode away, that would hinder us and our children and grandchildren from maintaining or *sustaining* productive agriculture well into the distant future.

One of the problems with this or any other definition of sustainability is that we do not know what future technologies our children will possess or what unforeseen problems

> While we should not try to refrain from utilizing resources, we should do so only on a scale that leaves room for future generations. We must consider our planet to be on loan from our children, rather than being a gift from our ancestors.
> **—Gro Brundtland, former prime minister of Norway, director of the World Health Organization**

they will have to address in order to obtain the resources that they need. In the case of agriculture, for example, agricultural practices are constantly changing to accommodate increasing demand for food, to redress environmental pollution caused by current and previous practices, and to take advantage of technological developments that improve yields. Unlike the simplicity of sustaining a single chord played on the piano, future sustainability of agriculture and other human enterprises will require composing new harmonies among ecology, economics, and technology that are more sustainable than the current practices.

The challenges of sustainability are not uniform across the globe. For the developed nations, the horse-drawn plow and the human-heaved hoe of the nineteenth century have long since been replaced by tractors, but these ancient methods are still widely used by peasant farmers in most tropical countries throughout the world. On the same day that the farmer in the Congo tills a small clearing in the forest with a traditional hoeing tool, a farmer in Iowa studies a computer screen inside his tractor. His computer is in constant contact with satellites in space, so that the computer can locate the position of the tractor very precisely on a digital map of the farmer's land. The computer instantaneously displays to the tractor operator the latest soil testing results and the latest data on fertilizer and pesticide application for the particular corner of the field where the tractor is located that moment.

How far into the future will the African farmer and her descendants be able to farm the same plot and feed their growing families using the hoe? Will the Iowa farmer and his descendants be able to use their increasingly sophisticated technologies to minimize their need for pesticides and fertilizers, thereby preventing groundwater pollution that might require wells to be closed for centuries? Must development in Africa and elsewhere first go through the same phase of overuse of fertilizers that occurred in most developed countries, or will agriculture in the developing world follow its own course, avoiding some of these mistakes? Can new technologies be developed that are appropriate for the people, cultures, soils, and economics of these very different parts of the world? And can technology continue to improve yields and at the same time reduce pollution sufficiently so that these practices will be sustainable?

Everyone seems to be in favor of sustainable development. After all, it would be hard to take the position, openly at least, that you prefer unsustainable development. Unfortunately, rewards for short-term profit taking are common in business today and are often associated with some sort of unsustainable exploitation of natural resources. A small army of researchers, extension agents, and development groups is seeking new approaches that might qualify as sustainable management of natural resources.

A Case Study of the Search for Sustainability

A Brazilian colleague of mine is one of the seekers of sustainable agriculture in his country. His name is Cássio

Pereira, he has a bachelor's degree in agronomy, and he is currently working on a master's degree. His innovative ideas, his rapport with the peasant farmers with whom he works, and his dedication to finding alternatives to their present practices of destroying forests to feed their families make him a hero in my mind. Cássio's approach was to start by carefully observing the farming practices of the rural peasant farmers (also called "small landholders" because each farmer owns only a relatively small plot of land) of the eastern Amazon Basin, to listen to the farmers' concerns, and to understand how they are adapting to the changing world around them.

These rural farmers have lived along the tributary rivers and streams of the Amazon for many generations. They are descendants of a mix of Portuguese colonists, African slaves, and Native Americans. Until about 1960, they relied entirely on the rivers as means of transportation in this mostly roadless area. They fished, hunted, gathered forest fruits and medicines, and they cleared small plots of forest to plant crops. They engaged in some trading and commerce along the river, but they were mostly subsistence farmers, meaning that they produced nearly everything they ate, and they ate nearly everything they produced. Their plots for growing rice, corn, and cassava were small and had to be abandoned a year or two after clearing because the soil would quickly become infertile and the fields would be overtaken with weeds. They would then clear and burn another small plot of forest to prepare for planting.

This type of agriculture has been called *slash and burn agriculture* or *shifting cultivation*, because they must cut and burn the forest and shift to a new plot for their crops every few years. As long as their population was sparse in this huge country, that type of agriculture might be considered sustainable. There was plenty of forest available to supply their modest needs of small fields. Once the fields were abandoned, the forests regrew and gradually reestablished the soil fertility. After several decades of forest regrowth and regeneration of soil fertility, the farmers, or more likely the sons and daughters or grandchildren of the farmers who did the first clearing, could return to these old fields and clear them again. As long as the ratio of available forest land to the number of people was high, this kind of rotation could be and was sustained for generations. Their lives may not have been as romantic or noble as some might be tempted to read into this description. These people had little access to education, health care, or any other modern amenities, but they generally got enough calories and protein to survive, and they passed on to the next generation the forests, rivers, and fields that were needed for their survival.

Everything changed in this region in the early 1960s when the government of Brazil built and paved a highway from the capital city of Brasília in the south to the Amazon port city of Belém, about 900 miles to the north. The highway is about ten or twenty miles from the river communities (as the macaw flies) where my friend Cássio now works, which is a long way in a roadless rain forest, and so they remained fairly isolated for a few more years. But before long, large cattle ranches sprung up along the main highway and along new side roads. Eventually, these side roads, which are passable only during the dry season, extended all the way to some of the river communities. Following the highways and roads were not only the cattle ranchers and land speculators but also poor landless people migrating from other overpopulated parts of Brazil in search of a place to clear some forest and plant some crops for sustenance. These migrating poor make claims on land by simply occupying it, hence they are often called *squatters*. These migrating poor have also been called the *shifted cultivators* (a play on the words "shifting cultivation") because not only do they shift their small plots from place to place, but they also often pick up and move to new regions. They have been "shifted" or victimized by the inequitable distribution of land and wealth in Brazil that has created a large landless class of migrating people in abject poverty.

The people that Cássio works with consider themselves natives of the area, as their great-grandparents lived there before them, but the current generation must respond to the introduction of roads, towns, markets, consumer goods, and the influx of people. One of their responses has been to grow more crops, which they can sell in the markets of the new towns. They use the cash to buy Western medicine, radios, watches, clothes, and other consumer temptations. They are starting to contribute to the Brazilian GNP. Growing more crops means making bigger fields by clearing more forest. At the same time, cattle ranchers and squatters have also cleared the surrounding forest, so there is less forested land available, and there are more people. This new situation is clearly not sustainable. Satellite images of the area taken in 1986 and 1991 show how much the forested area dwindled in five years. At the rate that the forest is being cleared in this particular area, it will be completely gone in thirty-five years.

1986
Deforestation near the Capim River, Brazil, as seen
from satellite images.

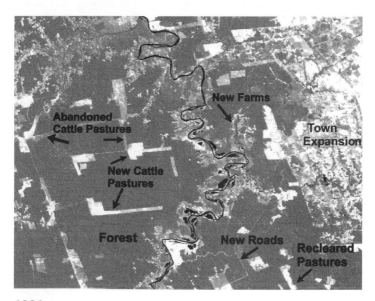

1991

Cássio also learned that the farmers prefer to cut *primary forests* (those that have never been completely cleared) rather than cutting *secondary forests* (those that are regrowing after having been cleared before). The farmers claim they get better yields of crops on land newly cleared from primary forest. The reason for this higher yield, Cássio discovered, was not that the soil was more fertile, but rather that the farmers could more easily control weeds in the fields created from primary forest than in fields created from secondary forests. In secondary forests, several weedy species of fast-growing plants first invade the area, and they or their seeds are still present in the soil when the secondary forest is recleared for planting, and so weeds are a big problem in fields created from secondary forests. Cássio also observed that the farmers were still using their traditional practice of planting corn, rice, and cassava in the same fields at the same time and without any orderly arrangements of rows.

Cássio convinced the town leaders to let him use some regrowing secondary forest land to create a small experimental farm. He cut down and burned the young regrowing forest, and he planted corn and cassava in one part of the farm, and he planted rice and cassava in a separate part of the farm. He showed the farmers how the tall corn stocks were shading the smaller rice plants in their fields and that by separating corn and rice into separate fields, they could get better yields of rice. After a few farmers tried this for themselves, they pointed out to Cássio another advantage of this system that he had never thought of. Most agronomists are trained to think about problems of weeds, insects, and plant diseases, but the farmers pointed out that birds had been feeding on their rice. The birds were flying in from the nearby forest, perching on the sturdy corn stocks, and reaching over to feed on the rice grains. By separating the rice and corn into separate fields, the birds no longer had a sturdy perch that they could use to get at the rice.

Cássio also arranged his plantings in rows that were designed for efficient hoeing of weeds. The traditional hoeing tool that the local farmers were using was fine for chopping out a few weeds at a time in fields created from primary forest, but it did not work well for weeding of row crops in fields created from secondary forests. So Cássio brought in some new hoes similar to the kind we use in our home gardens. He demonstrated that when the plants were properly planted in rows and with appropriate spacing for efficient hoeing with a good hoe, he could control weeds on the secondary forest land as well as the farmers were doing on the land cleared from primary forest. In fact, Cássio's crop yields on his experimental farm, using only his own hoe, ideas, and sweat, were better than the average yields that the farmers obtained.

Another innovation was to try several varieties of corn, rice, and cassava that might be better adapted to the acidic, nutrient-poor soil of this region than were the varieties that the farmers were using. Finally, Cássio planted trees in the field before the crops were ready to harvest. That gave the trees a head start without damaging the crops. The trees are members of the legume family, and they have the ability to take nitrogen, an essential plant nutrient, from the air rather than from the soil, which is an advantage in these nutrient-poor soils. These trees also have roots that go deeper into the soil than do corn and rice, and so they are able to obtain nutrients from a larger

volume of soil. The trees grow very fast and they accumulate nutrients in their leaves and branches as they grow.
Cássio let them grow only three years, at which time they
were 15–20 feet tall. He then cut the trees, burned the
site, and replanted corn, rice, and cassava. By using these
fast-growing, nutrient-accumulating trees, Cássio is hoping
to accelerate the recovery rate of the site so that another
crop can be grown in only three years, as opposed to the
several decades of natural secondary forest regeneration
that were required in the days of the farmers' parents and
grandparents. We will see if the yield from Cássio's second
crop rotation is as good as it was the first time around—
that is, whether his farming techniques are *sustainable*.
Actually, several crop rotations will be necessary to prove
sustainability.

You might be wondering why the farmers were not
planting in rows and using good hoes. Remember, however, that they used to have lots of land, and their crops
only supplemented the food they got from fishing, hunting, and fruit gathering. They did not need superefficient
cropping systems in those days. The fields created from
primary forests often still had several large logs crisscross-
ing the fields, which made neatly arranged rows of crops
rather impractical. They are now trying to produce more
crops for sale, they now have less forest to rely upon for future clearing and for hunting and fruit gathering, and so
they now need a more efficient cropping system.

Cássio's innovations are relatively simple, affordable, and
ready to be put into practice. His system is more intensive
than was the traditional approach, although it maintains
some of the traditional practices of rotating the land and
interplanting cassava with the other crops. He has the

farmers directly involved in the work, and information and experience is being exchanged both ways. Cássio's work is one hopeful example of many efforts that are in progress in South and Central America, Africa, and Asia to develop agricultural practices that accommodate traditional practices to the extent possible and to use modern technology where appropriate, while also responding to the inescapable economic and social changes occurring throughout the world.

The Diversity of Sustainable Agriculture

Although Cássio's innovations may help the plight of the small landholders of the region, who are poor but not destitute, these agricultural innovations will probably not be of much help to the squatters—the so-called shifted cultivators. Norman Myers, who coined that term, estimated that two-thirds of tropical deforestation is caused by shifted cultivators, who number in the hundreds of millions of persons worldwide. I am tempted to quibble over this estimate, as I think that Myers has underestimated the equally important destruction by commercial logging, but I agree with his main point that the landless migrating poor are an enormous problem that has devastating social, economic, and environmental implications. People who do not own land have little reason to be good stewards of that land. And most importantly, people who are so poor that they are uncertain of their next meal are too preoccupied with surviving in the short term to be seriously concerned about long-term sustainability. The most clever technological developments and the most innovative ideas concerning sustainability will not overcome the destruction that can be caused by poverty-driven desperation. Sustainable agriculture and sustainable development for the shifted cultivators will require difficult social decisions that may include changing land ownership rights or providing some other alternative for these people to find livelihoods.

Sustainable agriculture is a hot topic not only for the small landholder in the tropics but also for the medium- and large-scale farmers using modern technology throughout the world. The computerized tractor with a global positioning system is an extreme example of how the most recent technology is being applied in order to match the appropriate amount of fertilizer to the appropriate patches of a farmer's field instead of the more common approach of broadcasting excess fertilizer everywhere. Another innovation is to get rid of the plow altogether, or at least to use it sparingly. Plowing (sometimes also called *tilling*) the soil usually results in erosion, loss of organic matter from the soil, and leaching of nutrients to streams and groundwater. Eventually, fertilizers must be used when the native fertility of the soil is spent, and the use of fertilizer leads to the water pollution problems discussed in Chapter 6. As the soil continues to erode away and the surface waters and groundwaters become increasingly contaminated, it becomes clear that this system of agriculture is not sustainable.

Many farmers have started using *no-till* or *conservation-tillage* procedures in which the soil is disturbed as little as possible. Special seeding machines insert the seeds to the appropriate depths without first digging up the soil. By not tilling the soil, there is less chance of losing the soil to erosion, as well as less decomposition of soil organic matter. In fact, the organic matter that makes a soil rich with nutrients has been shown to gradually increase under conservation-tillage practices.

As in all things in life, there are trade-offs that must be considered with this new approach. One of the primary purposes of plowing is to control weeds. Just as my friend Cássio devoted a lot of his efforts to figuring out how best to control weeds with a hoe in slash-and-burn agriculture, weeds must also be controlled by farmers using the most modern technology. When tilling the soil is minimized, weeds can become a problem. One response is to use more herbicide, which increases the risk of contaminating streams, lakes, and the groundwater. Finding the right balance between tilling the soil to control weeds, not tilling the soil to prevent erosion and to build up organic matter, and applying herbicides as benignly as possible is now the focus of research in several universities across the United States, Europe, Australia, and in developing countries. Adoption of conservation-tillage practices by farmers occurs when all of the ingredients converge, including proven ideas and technology, effective dissemination of information, and economic incentives.

Despite tremendous differences in technologies, farmers throughout the world share an absolute dependence upon soil and climate, which constrain crop productivity. They also cannot escape the socioeconomic forces that influence human decisionmaking. Any attempt to achieve sustainable agricultural practices that focuses solely on socioeconomic factors while ignoring climate and soil, or vice versa, is doomed to failure.

Truly sustainable agriculture has probably existed during only a few brief periods in a few places. The example of the subsistence farmer-fisher-hunter of the eastern Amazon before 1960 may qualify as one example of sustainable agriculture. The Amish farmers of Ohio and Pennsylvania may qualify as another example. The other examples are probably limited to times and places where human populations were sparse, their technologies were crude, and their lives were hard. Since the advent of modern technology with its machinery, fertilizers, and pesticides, our agriculture has been extremely productive and has fed us reasonably well, but it has degraded soils and polluted groundwater and hence cannot qualify as sustainable. We have been feasting on borrowed time, waiting for the soil degradation problem and/or the groundwater pollution problem to reach crisis proportion. That need not happen if we move toward technologies that we think are closer to the ideal of sustainability. With growing populations and changing technologies, we may never pin down exactly what is sustainable at any particular place and point in time, but our concept is clear enough to identify the general direction that we must head.

ECOLOGICAL ECONOMICS AND SUSTAINABILITY

The discussion of ecological economics in previous chapters focused on "getting the prices right" for natural resources and "finding the right discount rates." Herman Daly, one of the leaders of ecological economics, has distinguished these tasks from the broader and more fundamental problem of the scale or magnitude of "how much" resource exploitation and use we should permit ourselves. He has noted that pricing and discounting fall under the purview of the subdiscipline of *microeconomics* but that ecological economics should also address the subdiscipline of *macroeconomics,* which deals with this question of "how much."

Daly has used the metaphor of loading a boat. Microeconomics can show how the correct prices and discount rates can produce an economic system that distributes the loading of the boat in the most optimal way so that the load will be balanced and the boat will not lurch to one side. Microeconomics will not, however, put a limit on how much can be loaded onto the boat. Left alone, microeconomics could distribute the load optimally, but it could *over*load the boat until it sinks. Similarly, we might use the tools of ecological microeconomics (the "right" prices and discount rates) to make the most efficient and rational uses of individual patches of forest resources, while neglecting to identify how much forest, for example, needs to remain intact to maintain a favorable climate and a reliable groundwater supply on regional and worldwide scales. Daly is working toward defining a macroeconomic system that recognizes the finite limit of the amount of forest, water, and air that can be exploited without destroying or fouling these resources. In a sense, Daly is defining how big the economic pyramid can be within the scope of the ecological pyramid shown at the end of Chapter 2, so that both economic and ecological systems can be sustained.

When the economic pyramid was small relative to the ecological one, such as in the days of the Wild West of the United States in the nineteenth century, we could have what Kenneth Boulding called a *cowboy economy.* The wild land was big enough to soak up our abuses, and the human economic system could ignore its localized detri-

Infinity is ended, and mankind is in a box;
The era of expanding man is running out of rocks;
A self-sustaining Spaceship Earth is shortly in the offing
And man must be its crew—or else the box will be his coffin.
                                    —Kenneth Boulding,
                        "The Ballad of Ecological Awareness"

mental effects on the environment. The cowboy could use the water, wood, and soil, leave his trash, and not recycle anything, because there were so few cowboys in such a big open space.

At the opposite extreme is the *spaceship economy,* in which every economic activity has a critical impact on our limited biological life support system, and vice versa. When planning to live on a small spaceship, every bit of energy consumed and waste produced must be carefully analyzed. If future population growth requires us to adopt a spaceship economy, everything will need to be recycled because nearly every drop of water and bit of energy will be needed to support the huge human enterprise.

At present, we are somewhere between these two extremes. We cannot be cowboys because the scope of our economic activities (the size of the economic pyramid) is so large that it has major impacts on the environment and we are seriously degrading our environmental life support system. We are not yet spacemen because we do not fully control all of the flows of energy and matter up the ecological pyramid. Instead, our economic system is like a bull in a china shop that is big enough to cause a lot of damage but that could and must be tamed. The role of ecological macroeconomics is to define how big the bull can be, while ecological microeconomics keeps him from breaking the china. Neoclassical macroeconomics does not recognize any such limits to economic growth, how big the bull can be, or how many cowboys or spacemen can be supported, because it assumes that technological advances can always substitute for spent or polluted natural resources. Ecological macroeconomics recognizes the potential for technological development but also recognizes the finite limit and nonsubstitutable nature of the most basic natural resources, such as soil, water, air, forests, oceans, and climate. Ecological macroeconomics is the economics of sustainability. We will return to this question of "limits to growth" in Chapter 8.

Links Between Sustainability and Equality

Sustainability shares several common traits with another holy grail, equality. Both are difficult to define clearly, although it is easy to give examples of what is unsustainable and what is inequality. Our past is full of tragic histories of unsustainable abuse of natural resources and inequality among our people. Neither sustainability nor equality has been fully achieved for long, if ever, and the future path to achieving them is not entirely clear. In his recent book, *The Ostrich Factor*, Garrett Hardin pointed out that equality is often used synonymously with equity. Despite Thomas Jefferson's famous statement that we "hold it to be self evident that all men are created equal," Hardin argued that no two people really are created equal (even genetically identical twins may experience differences in the womb that affect them differently later in life). Instead,

> Natural philosophy has brought into clear relief the following paradox of human existence. The drive toward perpetual expansion—or personal freedom—is basic to the human spirit. But to sustain it we need the most delicate, knowing stewardship of the living world that can be devised. Expansion and stewardship may appear at first to be conflicting goals, but they are not. The depth of the conservation ethic will be measured by the extent to which each of the two approaches to nature is used to reshape and reinforce the other. The paradox can be resolved by changing its premises into forms more suited to ultimate survival, by which I mean protection of the human spirit.
>
> **—Edward O. Wilson**

the modern day interpretation of the American Declaration of Independence is that all people, men and women, should be treated equitably by the law and that all children should have equal opportunities for obtaining a good education. Not everyone agrees with this definition of equality or with these goals, and in any case, they have not been fully achieved.

If we extend this concept of equality, however well or poorly defined, to equal opportunities among generations, then equality and sustainability become inexorably linked. For future generations to have equal opportunity to obtain wealth and well-being from soils, fresh waters, forests, oceans, and climate, we must not degrade these natural resources to the point that future generations can no longer effectively use them. Hence, if we want our children and grandchildren to have opportunities *equal* to ours, we must find ways to use our natural resources *sustainably.* Of course, those future generations may have more advanced technologies that will allow them to clean up dirty air and water and to make still more efficient use of the remaining soil, but we do not know how fast those technologies will develop. If they could speak up now, future generations would presumably want us to apply the precautionary principle rather than fall into Custer's folly of assuming that technological developments will arrive in the nick of time. We do not know exactly how much destruction and pollution we can get away with without coopting the opportunities of future generations, so we cannot define what is sustainable with much precision. Nevertheless, failure to unambiguously define what is sustainable is not grounds to dismiss the concept but rather is evidence that, as with equality, we have not fully committed ourselves to achieving it.

8 🌿

Fill the Earth and Conquer It, but Keep Two of Each Species

Can Both Imperatives Be Achieved?

A MEXICAN PEASANT WOMAN knew both her biblical lessons and the modern state of the world when asked if she saw a conflict between her use of family planning to limit the number of her children and God's admonition to "be fruitful and multiply and fill the earth and conquer it." "We've done that already," she replied. There can be no doubt that with about 6 billion people on the planet today and with at least 8 to 10 billion projected for the end of the twenty-first century, we have done well as a species. We have been fruitful, we have multiplied profusely, we have quite literally filled the earth with our numbers, and our technology has conquered many of the forces of nature.

Along the way to our dominance, however, the Old Testament tells how God became disappointed in what humans had done and He wanted to start over with a clean slate. When God instructed Noah to load up two of each species of animal onto his ark, there were so few people on the earth and their technologies were so crude that they probably had not destroyed much of the habitat of the wild animals. Noah presumably had little trouble finding a pair of each species. If God were similarly disturbed with our current behavior and had the inclination to try the forty-day flood again, a modern-day Noah would have to reply that finding a pair of each species can no longer be done—some of the species have since been lost forever and can never be loaded back onto the ark. Some were lost because of man-made floods when dams were built and reservoirs were filled to control floods, to irrigate farmland, and to provide a stable supply of drinking water for large urban populations. Many other species perished through just the opposite process, as wetlands were drained to produce solid land more suitable for human habitation. Other species were lost because forests were cleared to make way for cattle, sheep, and other domesticated animals of the growing human population. Still more were lost when native prairies were plowed to produce abundant grain to feed our many mouths. The same activities that have allowed us to be fruitful and to multiply and the same technologies that have allowed us to conquer the earth have also resulted in the loss of many of our former fellow passengers on the ark. The abundant diversity of animal, plant, and microbial species is often called *biodiversity,* and this diversity has been reduced by human-caused extinctions.

Today, most of these extinctions are occurring in tropical regions because of destruction of habitat. Although many types of habitats are being destroyed, the most rampant loss is currently in tropical rain forests. Average annual deforestation in the Brazilian Amazon Basin alone is about 1.5 million hectares (a hectare equals about 2.5 acres). For all of Latin America and the Caribbean, the rate of deforestation is over 7 million hectares per year, while less than half a million hectares of forest are being planted annually. Similarly, in Africa, over 4 million hectares per year are being deforested, while less than a tenth of a million hectares of forest are being planted. Southeastern Asia does better in the planting category, with about 2 million hectares planted, while about 4 million hectares are cut each year. However, about one-third of the plantings fail, and even those that do succeed in producing trees do not have the same richness of diversity of plants and animals as the forests that they replace. Farms and ranches are often abandoned after they lose their agricultural productivity, and secondary forests start to regrow. In most cases, however, these young forests are cut again after a few years, so that mature forests are generally not regrowing in many tropical regions. In sum, an area equivalent to the state of Michigan is being deforested each year in the tropics, whereas successful tropical reforestation covers only about one-tenth that area and is roughly equivalent to the area of Hawaii. Clearly, these rates of deforestation are not sustainable.

God's instructions to Noah have been interpreted by some as a moral imperative that we humans have a responsibility to protect, preserve, and help sustain our fellow species. If so, we have certainly failed, and our current actions are becoming more immoral with every passing year as habitat is destroyed and the number of extinctions

> People will conserve land and species fiercely if they foresee a material gain for themselves, their kin, and their tribe. By this economic measure alone, the diversity of species is one of Earth's most important resources. It is also the least utilized. We have come to depend completely on less than 1 percent of living species for our existence, with the remainder waiting untested and fallow.
>
> **—Edward O. Wilson**

climbs. Moral arguments, however, do not lend themselves to meaningful debate—either you accept them or reject them.

ECONOMICALLY VALUABLE PHARMACEUTICALS AND BIODIVERSITY

If you prefer more practical justification for preserving species, such as reducing human suffering and extending human life expectancy, you can find any number of examples of obscure little creatures that have turned out to produce human life-saving drugs. Virtually none of these microbes, plants, and animals appeared to have any practical value to humans until a unique chemical component of their bodies was discovered that proved valuable to medicine. They were always part of the ecologist's pyramid but did not become part of the economist's pyramid until a marketable pharmaceutical made from their tissues or made from their genes was discovered.

One of the more recent examples is that baboons appear to be unaffected by the AIDS virus. Research is being conducted to find whether injections of baboon bone marrow into human AIDS patients could allow them to recover some of their immune system's capability to fight off diseases. Such an advance in the effort to treat and cure AIDS would not be possible, however, if the baboons themselves no longer existed because their forest habitat had been cut down. So far, this medical technology has not succeeded, so baboons do not yet have much, if any, value in the economist's pyramid, and their forest habitat is still being cut.

The justification for the value of species based on a future pharmaceutical potential is fine, but I have my doubts that it is strong enough to save many species. We do not seem to appreciate the sources of the drugs that we already have. When you get a prescription filled at the pharmacy, you do not think about the little bug that created the drug that is going to make you feel better. Pills come in little bottles that bear no resemblance to the bacteria, fungi, plants, and animals that created the drug in the first place. Most pharmaceuticals are now synthesized by modern industrial processes, but nearly all of them have been copied from a chemical that was originally discovered in nature, produced by some creature in order to protect itself from another creature. Virtually every antibiotic was originally found in nature, but how much credit do we give the environment today for developing these life-saving drugs? If the environment is currently under attack when it has already yielded for us the most revolutionary drugs in medical history, why would finding a new cure for cancer or AIDS necessarily make us appreciate the environment any more? The new drug would quickly be put into pills and the little creatures would still be left to fend for themselves as we continue to conquer the earth.

There have been a few notable successful cases of pharmaceutical companies investing in the protection of biological preserves in exchange for the right to share the profits of pharmaceuticals derived from creatures found in those preserves. To the extent that these experimental deals prove profitable to the pharmaceutical companies and the local governments and people, they could succeed in preserving the biological diversity of some areas. I suspect, however, that these deals will be limited to relatively small areas and that pills will not be the salvation of biological diversity. One might think that the untapped potential for finding new cures for human diseases among the amazing diversity of chemicals produced by nature's creatures should be adequate justification for preserving their habitat. But as we saw in Chapter 4 about discounting, future potential seldom competes well in economic calculations with the immediate profits to be had from cutting forests.

DIVERSITY AS AN ESSENTIAL RESOURCE

The rationale that I prefer for preserving a diverse abundance of plant and animal species is that we depend on these creatures every day for our well-being. This is a hard argument to sell, I know, because one does not think of periwinkles, snail darters, or baboons as crucial to one's daily well-being. Just as we are separated from the importance of soil by the convenience of our supermarkets, seldom think about how the ozone high in the stratosphere protects us from harmful UV radiation, are generally oblivious of the warmth provided by greenhouse gases in the air, and take for granted that potable water will always flow

> The last word in ignorance is the man who says of an animal or plant: "What good is it?" If the land mechanism as a whole is good, then every part is good, whether we understand it or not. If the biota, in the course of aeons, has built something we like but do not understand, then who but a fool would discard seemingly useless parts? To keep every cog and wheel is the first precaution of intelligent tinkering.
>
> In human history we have learned (I hope) that the conqueror role is eventually self-defeating. Why? Because it is implicit in such a role that the conqueror knows, ex cathedra, just what makes the community clock tick, and just what and who is valuable, and what and who is worthless, in community life. It always turns out that he knows neither, and this is why his conquests eventually defeat themselves.
>
> **—Aldo Leopold**

from our faucets, we also seldom think about our dependence on the existence of a diversity of species of plants and animals throughout the world. Nevertheless, our need for a diverse flora and fauna is no less important, if it is more obscure, than our reliance on these other essential natural resources.

Paul Ehrlich has offered the best metaphor for the importance of plants and animals. The next time that you ride in an airplane, look out the window and try counting the rivets in the plane's wing. What if a couple of them were missing? You probably would not be too alarmed because, after all, there must be hundreds of rivets holding that wing together. Likewise, losing a few species among the millions that exist probably will not matter much. What if a dozen rivets were missing? That might get worrisome. Two dozen missing rivets? Maybe you should ask the flight attendant to point it out to the pilot.

How many species of plants and animals do we need to keep spaceship Earth together? We have already lost quite a few and we have not crashed yet, but how many more can we lose? If you single out a particular species, like the tiny snail darter that held up the construction of a big dam in Tennessee, you might argue that one is expendable. But how many times can you let one or two go before the strength of our craft and its overall functioning is at risk?

Experts in taxonomy estimate that there could be anywhere from 2 million to 100 million species on earth, and the best guess is about 7 million. That is a lot of rivets, but, then, the earth is big and its ecosystems are complex, and we do not know how many of these species we *need*. Of these many millions of species, we have names for about 1.4 million. Researchers are identifying only about 10,000 new species every year, so we have a long way to go to identify them all. Of the 270,000 plant species estimated to exist, scientists have identified about 34,000 known species of plants that are "at risk" of extinction because their habitats have been mostly destroyed to make room for humans. Many more species have not yet been identified by taxonomists, but, without a doubt, a great number of them are equally at risk.

Extinction is also a natural process, and many species have evolved and gone extinct without human interference. But the speed with which species are now going extinct because of human destruction of their habitat is almost unprecedented. A sudden burst of high extinction rates that marked the demise of the dinosaurs occurred some 65 million years ago, when a giant meteorite is

> What event likely to happen during the next few years will our descendants most regret? Everyone agrees . . . that the worst thing possible is global nuclear war. . . . With that terrible truism acknowledged, it must be added that if no country pulls the trigger the worst thing that will probably happen—in fact is already well underway—is not energy depletion, economic collapse, conventional war, or even the expansion of totalitarian governments. As tragic as these catastrophes would be for us, they can be repaired within a few generations. The one process now going on that will take millions of years to correct is the loss of genetic and species diversity by the destruction of natural habitats. This is the folly our descendants are least likely to forgive us.
>
> **—Edward O. Wilson**

thought to have struck the earth, ejecting debris into the atmosphere that suddenly and drastically changed the climate. There have been five major extinction events in the geologic record going back hundreds of millions of years, and they were presumably caused by catastrophic events such as meteorites and relatively sudden climatic changes. In the view of many experts, we now appear to be starting the sixth major extinction event in the earth's history, and this one cannot be blamed on a big rock falling from the heavens. This round of extinctions is being caused by us.

Because the exact number of existing species is unknown, and therefore the rate at which they are going extinct is not known with great accuracy, there is plenty of room for debate and for twisting around the numbers. In a recent book, presented as a debate between an environmentalist and a neoclassical economist, Norman Myers and the late Julian Simon argued about just how many species have actually gone extinct. Simon, the economist, argued that less than 100 extinctions could be documented (he was wrong; the actual number was 584 documented plant and animal extinctions as of 1992). Myers, the ecologist, used scientific theory to estimate that thousands or tens of thousands of species become extinct when large sections of tropical rain forest are cut down. In one sense, both men were correct—only a few hundred species for which we have names and descriptions can actually be documented to have gone extinct—but we also have very good scientific justification to believe that many more species that have not yet been identified or named, and hence cannot be documented or counted, disappear when their habitats are destroyed.

I hesitate to cite statistics like these about the number of species already extinct or in grave danger of soon becoming extinct, because counting species misses the main point. The number of species going extinct is a convenient yardstick with which to measure loss of diversity of life, but it is not really what matters most. Counting species is not like counting beans or counting dollars. Not only do species and dollars not equate, but all dollars are alike, whereas there is diversity both among and within species. The biodiversity discussion too often ignores the importance of diversity within species.

To illustrate this point, let us take human beings and beech trees as examples. Several tribes of American Indians have become extinct since Europeans invaded North and South America. The human species has certainly not become extinct, but we have lost some of the diversity that once existed within our species. The extinct Indian tribes died largely because they were susceptible to European diseases, such as tuberculosis and measles. Yet perhaps they would have been resistant to AIDS or to the ebola virus. Some people have genetic resistance to these diseases, and perhaps a race or tribe could be more resistant to one disease just as they are more susceptible to others. In addition to their cultures, which we cannot recreate, perhaps these Indian populations had some genetic endowment that would have helped the human race cope with an unforeseen future.

Similarly, a beech tree growing in Virginia has a different set of genes than does a Canadian beech tree. The Virginian beech tree, which is adapted to a warmer climate than the Canadian beech tree, may have genes that will prove adaptive when the earth warms because of our emissions of greenhouse gases. If only Canada saves its habitat for beech trees and Virginia does not, the species will not go extinct, but important diversity of life will have been lost. Indeed, many beech forests in New England and Canada have recently been attacked by a bark fungus. We do not yet know why this disease is currently spreading, but if the trees have become susceptible to this disease because they have been weakened by the warming that has already occurred, then the genetically diverse populations of beech from further south may become important for the survival of this species. Something similar has already happened in Sweden and Norway, where foresters are planting pine seeds that come from further south in Denmark and Germany. The plants from the southern seed source grow better in Sweden and Norway today because the climate is now warmer there, whereas the native populations of pine trees that evolved and once thrived in the previously cooler climate of Sweden and Norway are no longer well adapted to growing there. Had the habitat and seed sources for pine trees in Denmark and Germany not been maintained, the forest industry and the GNPs of Sweden and Norway would be much worse off now.

We have already discussed many of the functions that forests provide us that we often take for granted and that are seldom included in the economist's pyramid. Forests help moderate the earth's climate, they help purify groundwater and stream water, and they provide us with extractable products such as wood, fruits, nuts, and game. A forest with fewer species or with less genetic diversity could probably provide most of these same services for many years before we noticed a difference. In the long run, however, for a forest to remain healthy it must be able to withstand everything that nature throws its way over the course of time. There are floods, droughts, fires, windstorms, insect outbreaks, viral diseases, and even changes in climate. The best way, indeed perhaps the only way, to deal with this constant barrage of challenging conditions is to have a lot of options. Each combination of genes offers an ability to adapt to a unique set of conditions. If there are lots of combinations of genes in the forest, then the forest will have great resiliency. The more genetically impoverished it becomes, the less it will be able to adapt to the next change.

Change is certain. Every forest and every habitat experiences change eventually and must adapt to these changes. Forests and many other types of ecosystems are not as "fragile" as they are often described in pro-environmental literature. Like the boy who cried wolf, environmentalists have hurt their own cause by overusing this word. There are a few places where the word is apt, like the very fragile crust of lichen and algae that grows on the surface of some desert soils. Once broken by the wheels of an all-terrain vehicle, these important players in the desert life cycle are destroyed and do not grow back for many decades. Most forests, in contrast, can withstand a lot of tough weather from Mother Nature and even some use and abuse from humans. The rain forests of the Amazon and of most regions of the world have been modified by the indigenous people who have lived within them for centuries. These people have planted some trees, encouraged others, and destroyed some too. There may not be such a thing as a natural or *virgin* forest completely void of human influence. For the most part, healthy forests can and do recover from moderate levels of disturbance by humans. But if we are to retain forests and to benefit from their essential services of moderating climate and purifying water, then we must protect their ability to adapt to change by keeping them genetically diverse. That means keeping forests and other habitats full of many native species of plants and animals and keeping a variety of genetic diversity within each species as well.

So how many rivets do we need to keep the aircraft together? How many species can we afford to lose? Not only are these questions unanswerable, but they are the wrong questions.

How to Protect Diversity—Think Big!

One limited approach to keep species from becoming extinct is to depend upon zoos and a few national parks and nature preserves. When we set aside parks, the plants and animals in those parks will be protected for a few decades. Eventually, however, that park will be hit by a fire, a flood, or an invasion of a foreign disease or predator. Whether natural or man-made, some big event will happen that will wipe out many of the plants and animals in that park. Or perhaps it will be a somewhat less sudden change, like the warming of the earth during the next century that eventually will change the climate of the park, leaving it unsuitable to the plants and animals that lived there when the park was first set aside as a nature preserve. If the park is an island in the midst of a landscape void of wild habitat, then the park's plants and animals will never recover from these inevitable changes.

The limitations of preserving species in parks that are essentially islands in a sea of developed land can be gleaned from what has happened on many real islands in the ocean. A species of snail recently went extinct on the island of Raiatea in the South Pacific. Another species of snail had accidentally been introduced there, and in order to control the introduced snail, a predator was also introduced. Unfortunately, the introduced predator preferred dining on the native species of snail rather than the unwanted intruder, and the native species was completely wiped out. Had this accident not occurred on an island, there might have been another population of the same species of snail somewhere else that could migrate or be introduced to the area once the nonnative predator was removed, but this species of snail exists nowhere else, and so it is gone forever. The same can happen to species that exist only in parks if those parks are essentially islands of green in a landscape of human development.

Although the efforts of groups that acquire land to protect it in parks and preserves, such as the Nature Conser-
vancy and Conservation International, are laudable, they
are not enough. Parks and biological preserves often help
preserve the last remnants of a nearly vanished habitat
type, but these isolated parcels of protected land are insufficient for preserving species or genetic diversity in perpetuity. Instead, we must think big—really big. Parks and biological preserves are unlikely to cover more than 10
percent of the land that humans would like to inhabit. To
be sustainable, these parks must be surrounded by substantial buffer areas where rural human communities can
make a living from the land, but where habitat also exists
for many of the same plant and animal species found in the
parks. These parks and surrounding areas of relatively light
human habitation should cover half of the landscape. The
other half can be devoted to intensively managed agriculture, forestry, industry, and cities.

This 50/50 split is, of course, rather arbitrary. We could
argue about whether the split should be 50/50, 60/40, or
even 70/30, but it cannot be 90/10. It will not be enough to
keep only 10 percent of the land as parks while the other
90 percent is converted to intensive agriculture and urbanization.

We will need all of the ingenuity that modern technology
can muster to confine the majority of human productivity—our farms, forest plantations, industries, cities, recreation areas—to as small an area as possible, while maintaining a lighter human impact on the remaining vast areas
of land. Humans will still influence all of the world—that is
unavoidable with 6–10 billion people throughout the

twenty-first century—but we must tread lightly on about half of the land.

The story in Chapter 7 of my friend Cássio searching for sustainability with his Amazonian farmers is only one example of how people can make a living in a mosaic of forested and agricultural land. High productivity is needed in the land that is devoted to agriculture. Where forests are maintained, people will live within them and will make use of the wood, fruit, medicines, and game. To be sure, some species of plants and animals and diversity within species have been lost and will be lost because these people are there, but forest habitat with an abundance of diversity of plants and animals can be compatible with human habitation. Economic and regulatory incentives will be needed to maintain a significant part of the land in forest.

No single successful recipe for this way of life has been found; the ways to make it work are as diverse as the cultures of people in all parts of the world. The best tools of ecological economics and neoclassical economics will be needed, in addition to recognizing the primacy of the ecological pyramid. In some places, improving existing roads might be justified to facilitate access to markets for fruits, nuts, medicines, and other nontimber products of forests. In other places, new ways of making the crops more productive on the areas already cleared of forest are needed. Probably in all cases, a clear title to the land for the forest dweller is necessary. Our research institutions have only recently begun to focus on this type of agriculture and forest management, but we also have the benefit of considerable knowledge from local traditions and folklore. It will be a great challenge to find the appropriate mix of modern technology, economics, science, and political constituency building with traditional culture and knowledge to pre-serve and protect the world's forests, while also convincing governments to keep out the land speculators and short-term profit seekers in the meantime.

Maintaining diversity in forests or in any other habitat is not a matter of counting the number of extinct species or somehow measuring the genetic diversity and then decree-ing that we must stop development of the human eco-nomic enterprise when extinctions reach a certain number. Not even the best tools of ecological economics and ecol-ogy and the most accurate cost-benefit analysis could allow us to play God by saying exactly how many species are enough to protect or how many can be allowed to go ex-tinct. Rather, we need a fundamental change in the way we use land for human livelihoods that is also compatible with maintaining the greatest possible diversity of plants and animals on a significantly large part of that land.

Saving the Amazon for the Amazonians

Is a goal of 50 percent habitat preservation feasible? We have already appropriated over 95 percent of the produc-tive temperate grasslands of Canada, the United States, and the former Soviet Union for tremendously productive modern agriculture. It is unlikely that this habitat could be restored to native grassland while still feeding the world's population, and so a goal of 50 percent being used lightly is unrealistic for temperate grasslands. On the other hand, about 80 percent of the Amazon forest is still largely intact. Current trends of logging and clearing are depleting the Amazon forest at the rate of about 0.5 percent per year, but it is not too late to find ways to protect this forest while also providing a livelihood for the people who live there and supporting the GNPs of Brazil, Peru, Ecuador, Colombia, Bolivia, and Venezuela.

The Brazilian government has recently declared a goal of preserving 10 percent of the Amazonian forest in parks, biological preserves, and Indian lands. Most of the parks exist only on paper, with little or no presence on the ground that would exclude loggers and squatters. The native American Indians are carrying out timber extraction and agriculture on some of their lands, but most of their preserves remain forested. Although many environmental groups have complained that the goal of 10 percent is not enough, at least the terms of the debate within Brazil have been shifted from "how much deforestation can be allowed" to "how much land must remain in forest." If 10 percent can indeed be set aside in government preserves, and another 40 percent or more can be maintained surrounding those parks in some kind of privately or publicly owned forest where forest management is mixed with agriculture, then the goal of keeping 50 percent of the Amazon Basin predominantly in forest could be achieved.

Would keeping 50 percent of the Amazon Basin in forest impede the economic development of Brazil, restrict growth of its GNP, and harm the future economic well-being of the average Brazilian living in the Amazon Basin? What would be the economic effect of "locking up" 10 percent of the Amazon as parks and restricting the types of development that could occur on another 40 percent of the region that would remain mostly in forest? My colleague Daniel Nepstad has turned the question around:

Amazonia is racing toward a fate that has befallen virtually every major forest formation in the world. A short-sighted, self-serving economic and political elite is rapidly mining forest wealth, stealing from Brazilian society the hope of a future in which the region's vast natural resources are managed as the basis of an expanding and enduring prosperity.

The myth that dominates the media and the minds of Amazon citizens is that environmental concerns are exogenous, counter-development and the luxury of the rich. The myth equates the conservation of natural resources with the loss of economic opportunities. It views the rampant expansion of roads, electrical grids, and waterways as the good and necessary prerequisites of economic progress.

The major impediment to the emergence of an environmental constituency in Amazonia is the absence of compelling, accessible, credible information describing the ways in which present and future environmental degradation impoverishes the lives of Amazonians. People become concerned about the future of their natural resources when they realize that their own health and well-being are tied to the health of the region's ecosystems. The Amazon forest fires of 1997 and 1998, for example, were of great immediate concern to those who breathed smoke for weeks on end. In a similar way, Amazonians must realize that their smoke-filled lungs, their malaria, the assassination of their rural leaders, their illiteracy, and their rural poverty are the highly predictable outcomes of the government's policy of rapid frontier

expansion—policy which has never been effectively
opened to public debate.

The present type of frontier expansion and its accompa-
nying deforestation are perpetuating rural poverty, not al-
leviating it. When land is cheap, it is often squandered. As
seen in Chapter 2, this was true in the cotton fields of
North Carolina in 1880, and it is also true today in the
Amazon Basin. After the forest is cleared and the native
fertility of the land is spent, the people who eked out a living on the cleared land then abandon it and move on to another newly opened frontier where they find more cheap
land. They are still poor, and they leave a trail of impoverished land behind them. When the government builds new
roads, still more land becomes available, and as the supply
of land increases to meet demand, the price stays low.

Predominantly two groups become wealthy from frontier expansion. The companies that extract valuable hard-
wood timber, often using government subsidized roads,
make a large, one-time profit. And the land speculators
who buy up the cheap abandoned land may eventually
make a profit if they can consolidate several landholdings
to create viable ranching operations. These large ranches
support relatively few cattle over large areas and employ
only a few ranch hands. The real economic benefits of this
model of development are concentrated in the hands of a
political and economic elite. At the same time, the natural
resource base that could provide present and future wealth
for larger numbers is being squandered. The fallacy of
Marie Antoinette economics is at work: The Brazilian
GNP increases as the wealthy Brazilians get wealthier, but
the peasants and the land that they rely upon for their eco-

nomic well-being become and remain impoverished. In the long run, the entire nation suffers environmentally, socially, and economically from this impoverishment, despite the short-term, one-time gain in GNP provided by deforestation. Development can and should occur, but it should follow a different model.

In what economic and landscape setting does it make sense for Amazonian farmers, for example, to shift to agricultural practices that allow them to get more food out of less land? As long as forest land is accessible, abundant, and cheap, the most logical way of surviving in the Amazon continues to be shifting cultivation by small landholders and extensive cattle pasturing by large ranchers, both of which result in widespread deforestation. Government road building lowers land values by opening up new frontiers, thereby increasing the supply of cheap land. To find an alternative model of development, Brazil and other countries must be willing to postpone the opening of new roads. Only when easily accessible land starts to become scarce, does the value of the land increase. As the value of land increases, so do the incentives to invest in infrastructure on the land, to intensify productivity on the land, and to manage it wisely. When there are investments of monetary and human capital on the land, then there is a chance that the natural capital might also be appreciated and managed wisely. Both the small landholder and the large cattle rancher can benefit from gradually accumulating investments in the land and from increasing land values.

It may seem counterintuitive that providing cheap land is not the answer to poverty, but experience shows that opening up frontiers, alone, does not alleviate poverty. Viable strategies are needed to allow people to stay put in one place and to make a decent living. Providing access to more and more cheap land by building new roads is counterproductive to developing needed strategies for sustainability. The pros and cons of these governmental policy decisions and their effects on various economic classes of the population have not been debated. Instead, the model of road building and frontier expansion is being blindly followed without debate among those who are most affected.

Those Who Live in Wooden Houses Should Not Throw Away Chain Saws

I live in a wooden house, and I am not opposed to cutting trees and to managing some of our forests for producing timber. Most of the logging occurring today in the tropics, however, is more like mining than forest management. The valuable trees are removed and the land is either abandoned or converted to temporary agriculture. Although there are many tropical forest tree species about which we know very little, we know enough about managing forests in general to make more prudent use of these forests. Forest management includes a role for chain saws, but not a reckless one. Managing some of the forests efficiently and sustainably for producing timber will be essential if other large areas of forest are to be managed for nontimber products, for light intensity of human use, for the clean water and air that they provide us, and for the habitat that they provide to an astoundingly diverse array of plants and animals.

We have conquered the earth in many respects, but we have not yet learned how to live on it in a way that will not degrade it. Our current approach of expanding frontiers and expanding the influence of our modern technology everywhere will not work. We need modern technology to keep us fed, but we cannot afford to lose the variety of life that it is replacing. In this experiment of human domination of the earth, we do not yet know if both imperatives—conquering the earth and preserving most of our fellow passengers on the ark—can be achieved. We can say that our current course of action will drive more and more species to extinction and will eventually undermine our own ability to sustain ourselves. A new approach is needed that combines our modern technology with a rational apportioning of the land to meet our intertwined ecological and economic needs. Those needs include food and homes for billions of people and they also include preserving the functions that can be provided only by vast areas of intact forest habitat with a diverse array of plants and animals.

9 🌿

May We Live in Interesting Times

Some Modest Proposals for Profound Changes

IN THE BEGINNING, God created Heaven and the Earth, and the Earth has been the same size ever since. Yet the number of people and the constructive and destructive power of their technology has grown leaps and bounds.

President Reagan had a popular vision of re-creating in the 1980s the "the good old days" of the Eisenhower era in the 1950s. Americans had confidence then in the technology that had won the war, crime was a manageable problem, there seemed to be ample resources to go around, and the environment was simply not an issue. Reagan's mistake, I think, was to ignore the fact that there were 2 billion more people in the world in 1980 than there were in 1950 (and another 1.5 billion have been added in the short time since then). The majority of those people are not starving (although probably 1 billion or so are malnourished), but they are all demanding a piece of that

185

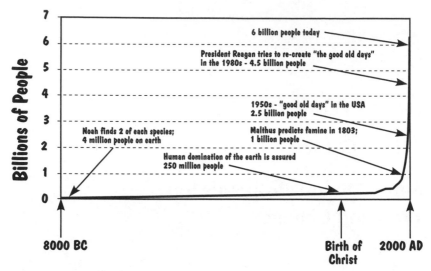

Human Population of the World

"better life" that Americans wanted in the 1950s and 1980s and continue to want today.

We cannot go back to the 1950s or to any other era. Human populations are growing and technology is advancing, both of which fundamentally change the way we live. At the same time, the resources of soil, water, air, forests, and oceans that supply the well-being of the growing human population and that absorb its garbage are either constant or are dwindling. Thanks to advancing technology, we can do more with less, and so we manage, sort of, with the growing population. At the same time, however, the human population and the garbage that it produces are growing like compound interest, and we live in an ever more complicated world. Although it is tempting to look back fondly at what may have been simpler times, the laws of technodynamics, presented in Chapter 6, ensure the inevitability that we will live under the Confucian curse of increasingly complicated and interesting times.

NEOCLASSICAL VERSUS ECOLOGICAL VIEWS OF POPULATION GROWTH

Concern over population growth is common today, but surprisingly, it is not universally recognized as a problem. In fact, neoclassical economic view is to the contrary. According to many neoclassical economists, each person is a resource, a source of potential ingenuity, a worker, and a consumer who will contribute to the GNP. The problems that arise from a growing population, such as the challenges of producing enough food and disposing of our chemical garbage, are seen by these economists as simply a greater stimulus and impetus to develop new technologies, which in the end will improve our lives further. If ingenuity and technology could substitute completely for natural resources like soil, groundwater, and the atmosphere, then according to this economic theory, there would be no limit to how many people the earth could support. By now, I hope I have convinced you that many natural resources are not substitutable and that this narrow economic view of how the world works is flawed and dangerous. Improved technology will be needed in this changing and ever more complicated world, as will the skills of economists, but we will always be part of the ecologist's pyramid that ultimately depends on irreplaceable natural resources.

The ecological macroeconomics of Herman Daly, discussed in Chapter 7, argues that the size of the economic pyramid is limited by the biophysical constrains of the ecological pyramid. If this is true, then we may choose between having a large human population with a low average prosperity per capita or a smaller population with a somewhat bigger piece of the "better life" pie for each person. By "better life," I mean not only the GNP per capita but also the nonmonetary values like uncrowded open space, wildlife habitat, clean air and water, and the assurance of a relatively stable climate.

A recent book by Joel Cohen, titled *How Many People Can the Earth Support?,* provides a fascinating account of the history of human population growth and the factors that have affected population growth in various cultures at various times. The question posed by the book's title, however, is never answered. I have not found a good answer to this question, and I doubt that even ecological macroeconomists like Herman Daly have a good answer either. As in the case of defining sustainable agriculture, the ever changing factor of technology makes the definition of a sustainable human population elusive.

I would argue, however, that how many people the earth can support is the wrong question to be asking. Rather, we should be asking if we already have enough people and perhaps too many people. If the neoclassical economic model is right—that each additional person is a valuable worker, consumer, contributor to the GNP, and stimulator of innovative technologies that can substitute for all natural resources— then we have nothing to worry about as the population grows. If the second law of technodynamics is correct—that the increasingly difficult challenges of consuming nonsubstitutable resources, providing food, and disposing of garbage for a rapidly expanding population leaves us and future generations with fewer options and more problems to resolve— then we already have too many people on the earth.

This is not to say that no one should have children. I rejoice in the memory of my son's birth, and I wish the same for all parents. We can celebrate the creation of life and, at the same time, recognize that the number of those creations should not exceed the number of deaths. I believe that encouragement of effective family planning can be achieved with ethically acceptable forms of education and persuasion. Although the heavy-handed forms of coercion used by the Chinese government to limit family size would not be acceptable in most Western cultures, a limit of two personal exemptions for children on our income taxes might be an acceptable and fairly effective incentive for many Western families. Why should the government reward people for having more than two children by giving them more tax deductions with each additional child? Ecological economics could help us find other culturally acceptable incentives to curb population growth.

Garrett Hardin has argued that discussion of human *rights*, including the right to bear children, should be accompanied by a discussion of *responsibilities*, such as a limit to family size. If we do not succeed with responsible and effective family planning, then I believe that the day will arrive when we will not have the ecological and economic resources to celebrate each new birth. Indeed, that day may be near, and some would argue that it has already arrived.

The previous chapters have chronicled the effects of increasing population on soils, forests, groundwater, and the atmosphere. What about the human social and economic condition—the "better life" that the current 6 billion people on the earth know about and want, but that a relatively small number of us enjoy? Commonly cited statistics on our progress toward that better life can be deceiving. Thanks to a worldwide effort to improve access to clean drinking water, 750 million people gained access to sanitation services during the 1980s, which sounds like great progress. However, about the same number of people were added to the world's population during the 1980s and so the number without good sanitation remained constant. In the early 1990s, progress in providing access to sanitation did not keep pace with population growth, and the number of people without adequate sanitation grew from 2.6 billion in 1990 to 2.9 billion in 1994.

The number of children enrolled in primary school increased by 50 million in the early 1990s, but the number of school-age children also increased by about the same number, so the number not able to attend school remained constant. The mortality rate of children under the age of five has gone down thanks to successful immunization programs, but more and more of those children are surviving only to grow up in abject poverty. In only the four years between 1990 and 1994, the number of people living in absolute poverty increased from 1.0 billion to 1.3 billion. More than 1 in 5 people on the earth are extremely poor. Improved technology is providing better sanitation, more educational opportunities, and better health care for many, but the improvements are barely keeping pace with population growth and in many cases fail to do so.

People living in abject poverty usually do not have the luxury to care about natural resources, as they worry instead about their next meal and obtaining their most basic needs. In the worst cases, their destitution leads to destruction of their environment. At the same time, however, our ability to create the wealth and economic resources that might enable those people to escape poverty will require that we use our natural resources as wisely as possible. Hence, we have the doubly challenging task of making prudent use of our soils, air, water, forests, and oceans while trying to meet the demands of an increasing number of impoverished people. The task would be much easier if there were fewer people on the earth.

Arguing about how many billions of people can be supported is a waste of time. Clearly, we do not need any more people or any more problems to provide stimuli, as the neoclassical economists argue, to spawn ingenuity and development of new technologies. Our considerable environmental, economic, and social problems already provide plenty of stimulus for technological development and the increased economic prosperity that technology might deliver. In fact, the greatest challenge for technology, innovation, and human creativity, and probably the most effective way to increase per capita prosperity and per capita "better life," is to slow the rate of population growth and maybe eventually decrease population.

Prospects for Slowing Population Growth

Like ongoing global warming, additional population growth during the next few decades is unavoidable at this point. There are already so many young women of childbearing age, or soon to be of childbearing age, that even if they had only two children each, more children would be born in the next few decades than the expected number of deaths. Also like global warming, this trend could eventually be reversed, and the sooner we do so the fewer problems we will have.

This all seems very logical and reasonable, but between 1980 and 1992 the official policy of the United States government was that population growth was good. The issue of family planning and population control got mixed up with an ideological struggle over abortion. No U.S. federal dollars could be used to support clinics or family planning programs anywhere in the world that also provided information on abortion. This political football continued to be tossed back and forth in the 1990s, as the U.S. Congress and the president battled over policy. The irony, of course, is that better family planning would reduce unwanted pregnancies and thereby reduce the demand for abortions, so one would think that the anti-abortion forces would have supported family planning. Most people who oppose abortion probably do support family planning, but many of the leaders of the pro-life movement have an agenda that defies this logic.

Controlling population growth is a topic that often engenders gloom and doom predictions, because it will not be easy to accomplish. But once again, if we acknowledge the problem and apply our considerable knowledge, skill, experience, and technology to the task, the population explosion can be brought under control. We know, for example, that one key component to effective family planning is to provide educational and employment opportunities to women and girls so that they have attractive alternatives to staying at home. When women have good opportunities for careers outside the home, they and their husbands often choose to have fewer children.

Demographers have observed that the rate of population growth has started to slow. Some demographers predict

that the earth's human population may level off somewhere between 8 and 10 billion people toward the end of
the twenty-first century. This prediction is based on some
very preliminary trends and is premised on the assumption
that the demographic transition that has slowed population
growth in industrialized countries in the past is starting to
occur and will continue to occur in developing countries.
Although the reasons seem to vary from country to country
depending on cultural attitudes toward marriage, sex, and
gender roles, most industrialized countries experienced
decreases in their fertility rates as their economies developed. Presumably, the same will happen or is already happening for countries like India and Mexico. Even if these
predictions come to pass, however, we should not be san-
guine about a global population of 8 to 10 billion people.
The problems and challenges associated with feeding and
disposing the garbage of a human population stabilized at
only 8 billion people would be sufficient to keep the lives
of our children and grandchildren *interesting*. Early signs
of a change in population trends, while encouraging,
should not be misconstrued as dampening the urgency to
slow population growth and to stabilize the population at
lower levels. Because population growth is the engine for
the forces that strain our ability to use soils, forests, water,
and oceans wisely, there is no time for complacency on this
issue or for falling into the fallacy of "not beating the wife
as much as before."

We have not even begun to tap the potential that technology has to offer in making contraception easier,
cheaper, safer, and less awkward. Because of threats of
boycotts and lawsuits by the pro-life movement and lack of
research support from government, the ingenuity and

profit motive of the private health industry to improve contraceptive devices has been stifled for the past twenty years or more. Surprisingly, no new developments have had a major impact on family planning since the birth control pill. If technologists can dream up shields in space to replace the natural ozone layer that keeps harmful ultraviolet radiation from penetrating the atmosphere, why, then, can we not find more simple technological fixes to shield egg from sperm? Technology alone will not solve the population problem, but it can make a major contribution if we allow reason to rule. In this case, our enemy is not greed for short-term profit as it is for imprudent exploitation of most natural resources, but rather our enemy is ignorance and ideology.

Are There Limits to Growth of the GNP?

Nothing will provoke a debate from a neoclassical economist faster than a statement that there are limits to economic growth. But perhaps this argument, like the one about how many people the earth can support, has also been misconstrued and should be redefined. When I think of limiting the size of the economic pyramid, I think of keeping most of human activity constrained to about 50 percent of the once forested land, as described in Chapter 8. We need to constrain the economic engine by the amount of forest that the ecological system needs to function. This constraint, however, is not the same as limiting economic growth outright. Rather, technological development on the 50 percent of the land devoted to human industry, agriculture, and urbanization can still allow us to increase our economic output and the GNP. One prominent economist, Theodore Panayotou, sees the greatest potential for economic growth to come from increased efficiency and innovation:

> Skeptics ask whether growth is possible if the full environmental costs of growth are paid for. The answer depends on the source of growth. If the growth is derived from appropriating other people's resources or shifting one's own costs onto others, it will not continue. If it is derived from increased efficiency and productivity, it will continue. In fact, empirical studies show that the most important sources of growth are increased efficiency and innovation resulting from accumulated knowledge.

On the land devoted primarily to human economic enterprises, where this increase in efficiency must take place, we must also constrain the activity such that the soils, atmosphere, and groundwater are not impaired for future generations. The tools of ecological economics will be vital if we are to devise ways to preserve these irreplaceable resources while still rewarding ingenuity, efficiency, and economic growth. In general, there must be constraints to economic growth, but not necessarily limits.

One important example of an absolute limit, however, is the fishing industry. It was once thought that the fish in the sea were as infinite as the stars in the sky, but nearly every commercial fishery has suffered serious decline during the twentieth century, many are threatened, and some are closed. The demand for fish and our technological capability of building bigger boats and ever more efficient trawling and purse-seine nets has outpaced the ability of fish to reproduce. Government subsidies for industrial fishing fleets are largely to blame. The only way for the fishery commons to be preserved is for the stakeholders, including the village fisherman and the giant fishing corporations, to find equitable ways to limit their catch and to agree to set aside fish breeding grounds as sanctuaries. In this case, technology has overstepped its usefulness and has permitted an overexploitation of a limited resource. Pollution of the oceans and climate change make matters worse. The oceans cannot be expanded, and technology cannot increase the wild fish populations. A limit to the food and the GNP that can be derived from fisheries has truly been reached.

Aquiculture may offer some opportunities for expansion, but the two economically largest forms of aquiculture, salmon and shrimp farming, actually make matters worse. The salmon and shrimp raised in aquiculture "farms" are fed other fish that are caught from the open sea, thus creating more, not less, pressure for overfishing, as well as creating a nutrient waste (garbage) problem downcurrent of the aquiculture farms. Consistent with the first and second laws of technodynamics, expanding aquiculture is creating new, more difficult problems of pollution of estuaries and nearshore marine ecosystems.

CONSTRAINING THE ECONOMIC PYRAMID: THE TOP-DOWN APPROACH

Even if we can overcome ignorance, desperation, greed, irrationality, ineptitude, and forgetfulness, the earth is already irreversibly committed to a larger human population, a warmer climate, degraded soils, polluted and depleted groundwater, and dwindling forests. But we are not necessarily destined to endure environmental catastrophe similar to the famine originally predicted by Malthus. Some environmental catastrophes have already occurred, like the loss of the Aral Sea, degradation of large areas of arable farmland, irreversible extinction of species, and collapse of fisheries. The second law of technodynamics holds that the future problems are bound to be bigger and more difficult to solve. Nevertheless, we know a lot about how the world works, both economically and ecologically. The past catastrophes and most of today's environmental problems were foreseeable and preventable with our existing knowledge and technology. If environmental problems are acknowledged, they can be dealt with effectively, and future problems can be averted. The first step is where this book began: recognizing that natural resources—soil, water, air, forests, and oceans—are still essential in this complicated, technological world.

Once the necessity of these natural resources is recognized, the urgency of several national and international initiatives becomes apparent:

1. Stop building new roads. This idea may not seem unusual for those who have been trying to protect the few remaining wilderness areas of North America from logging and mining exploitation, but it is a revolutionary idea in the developing world. It is widely assumed that building new roads will open up frontiers of development that will lead to new economic prosperity. Many governments and international agencies finance the construction of roads, erroneously believing that this will improve the ability of their citizens to make a good living. Contrary to common expectations, however, the landless people who migrate to these newly opened areas usually remain locked in poverty, while a small political and economic elite of the country reap

nearly all of the economic benefit of the frontier expansion.
Truly sustainable development that will bring large numbers
of people out of poverty will require intelligent intensifica-
tion of agricultural and industrial productivity in the areas
already accessible by existing roads. Frontier expansion is
usually a false promise of economic prosperity.

2. Eliminate tax deductions for more than two children
in countries that have income taxes. Explore other financial and cultural incentives for family planning that are re-
spectful of the norms of each culture while also effectively
addressing the responsibilities of parents to limit family
sizes. Extend financial incentives and remove threats of
lawsuits to the health industry for development of new
technologies of contraception. Disentangle family planning from the politics of abortion.

3. Reduce taxes on income and increase taxes on consumption. For the most part, innovation, entrepreneur-
ship, and generation of income should be encouraged, not
discouraged by taxes. Likewise, wasteful consumption, especially consumption that causes pollution, should be discouraged by taxes. Taxes on energy consumption and the
use of pesticides, for example, will encourage conservation
and discourage pollution. Combining lower taxes on income with higher taxes on consumption will stimulate
profit-driven innovation that makes wiser use of natural resources at affordable prices. Because consumption taxes
are more regressive than income taxes and place a large
burden on poor people, companion policies will be needed
to cushion this burden on the poor.

4. Eliminate governmental subsidies, such as water projects that hinder the market forces that would otherwise
promote water conservation, more prudent use of ground-

water and surface water, and prevent soil salinization. Similarly, eliminate government subsidies of grazing and mining on public lands, which result in overgrazing, disturbance, and degradation of soils. The worldwide governmental subsidies of agricultural production, energy use, road transportation, water consumption, and commercial fishing that create perverse incentives for environmental degradation have been estimated at about $1 trillion of taxpayer money per year. In comparison, a comprehensive conservation program would probably cost about one-third that much.

In many countries, weak laws for protecting the environment from pollution and weak enforcement of existing laws are, in effect, another form of government subsidy to the polluters who do not have to pay the costs of their pollution. When pollution is ignored and not included in production costs, agricultural and industrial products are often more competitive in the international marketplace. Ignoring pollution as an economic externality may provide a short-term economic benefit to the exporting country and increase its GNP, but degradation of the environment will eventually undermine that nation's economic prosperity, impoverishing the environment, the country, and its citizens. Hence, thousands of people gathered at the World Trade Organization meeting in Seattle in 1999 to protest the free trade policies that ignore the social and economic costs of pollution and human rights abuses. Just as I argue that economic policies cannot be isolated from the larger ecological support system, the rebuttal to unfettered free trade policies argues that trade cannot be isolated from the social and environmental milieu in which it occurs. The editors of *Nature* were even more emphatic: "Free trade, whatever certain economists, politicians, and professional trade negotiators at the WTO may believe, is not more important than the future of life on our planet." International trade agreements must address environmental concerns, so that member nations cannot achieve unfair trade advantages from subsidizing, permitting, encouraging, or failing to discourage environmentally destructive enterprises.

5. Support governmental and nongovernmental extension efforts to encourage farmers in both industrialized and developing countries to make more efficient use of their land while also conserving soil and using only essential quantities of fertilizers and pesticides. Develop ecological economic incentives for soil and water conservation and integrated pest management.

6. Eliminate governmental subsidies for industrial fisheries and foster agreements among nations and among and within communities to protect the fisheries commons.

7. Ratify and enforce the binding international agreement reached at Kyoto, Japan, in 1997 for reducing greenhouse gas emissions. Using the tools of ecological economics, encourage development of technologies that will move us away from economies based on coal, oil, and gas. Carbon emissions trading, carbon or energy taxes, and tax incentives for renewable energy are among the leading possibilities for effective action.

8. Negotiate international agreements on maintaining forest cover and managing forests to maximize genetic diversity of plants and animals. Find national and local policies that provide economic incentives for landowners to keep significant portions of their land under forest cover.

Notice that there are as many items in this list about eliminating current government actions as there are calling for new government initiatives. Government is part of the solution, but it has also been a large part of the problem by creating perverse incentives for environmental destruction that benefit only a small political and economic elite. Although I have been critical of neoclassical economists throughout this book, many of them will approve of the items on this list that call for eliminating harmful governmental subsidies and governmental interference with market forces. On the other hand, government can also play a positive role by creating the right kind of economic incentives, which is where the new field of ecological economics fits in.

This list is by no means complete, but I prefer not to overemphasize national and international policies at the expense of slighting the importance of personal choices of individuals. Lists of governmental policy actions let most readers off the hook, because few readers are in a position to have much effect on these large-scale, top-down environmental policy approaches. Profound change in the way we think about economics and ecology will require a bottom-up approach in which every reader can participate.

Constraining the Economic Pyramid: The Bottom-Up Approach

As with identifying sustainability, it is easier to say what not to do than it is to say what you should do. After reading this book, I hope you will not follow the admonition of the winning poker player as he lays down his hand and proclaims, "Read 'em and weep!" My intent is not to engender gloomy, doomy hand-wringing and weeping. Instead, here is a modest list of positive steps, from easy-to-do to hard-to-do, that any reader can attempt.

Lend this book.   Perhaps it is not as catchy an admonition as *Steal This Book,* and it won't contribute to the GNP as much as would *Buy This Book,* but it will help generate debate and discussion. To help this book go beyond preaching to the choir, please lend it to someone who is unlikely to buy literature on science, economics, and the environment.

Analyze your habits of using electricity, heating, water, automobiles, consumer goods, and food.   Buying somewhat more expensive organically grown produce, for example, not only provides some peace of mind about what you are eating, but more importantly, also supports the development of technologies and markets that rely less on polluting pesticides.

Sometimes the analysis must go beyond the obvious. Although recycling is obviously commendable, don't leave the engine running while dropping off glass at the recycling center. Switching to an energy-efficient compact fluorescent lightbulb may not save the world (although it could make a difference to global warming if millions of people also switched), but it definitely saves you money on your electric bill, and its symbolic significance should not be underestimated. If turning on a lightbulb is symbolic for having a bright idea, then turning on an energy-efficient lightbulb can symbolize a vision of where our comfort and well-being come from and how we can use those natural resources efficiently and wisely. Don't stop with the lightbulb. A booklet by the Alliance to Save Energy (ASE) gives numerous tips for conserving energy around the home and workplace (http://www.ase.org).

Do you ever wonder whether it is better for the environment to send disposable diapers to the garbage dump or to consume lots of energy washing reusable diapers with hot water? What about choosing paper or plastic bags at the grocery store? A book by the Union of Concerned Scientists (UCS), *Consumer's Guide to Effective Environmental Choices* (http://ftp.ucsusa.org), analyzes these choices and dozens of others that we face in everyday life. When these choices are made by millions of people every day, they often collectively have profound effects on the environment.

In the case of paper versus plastic, the choice is trivial, and a far more important consideration is what kind of car you drove to the grocery store. If you drove a gas guzzling sport utility vehicle (SUV), there is no way you can make amends by piddling around with the environmental impact of paper versus plastic! Save the SUV for hauling cargo over dirt roads, and drive something that gets better mileage to the grocery store. With regard to the disposable or washable diapers, I'll leave you in suspense, in hope that you will be motivated to seek out this book and read also about the other numerous informed choices that we all could be making. Like the grocery bags, several consumer choices have only trivial effects on the environment, whereas many other choices about cars, food, home construction, and household use of heating, air conditioning, lighting, appliances, water, and sewage are very important.

An excellent real-life example of understanding the difference between trivial and important effects of consumer choices is given by the authors of the UCS book. They recount how a church group tried to do the right thing by eliminating paper and plastic cups used at church functions. They proposed to spend several hundred dollars to buy a dishwasher and reusable coffee mugs and glasses. The UCS consultants pointed out, however, that limited church funds would be much more effectively used to invest in increased insulation around drafty doors and windows. Cutting down on the energy needed for heating would produce a much larger positive effect on the environment by reducing greenhouse gas emissions and other air pollutants than would eliminating paper and plastic cups from the garbage. The investment in insulation would also have a payback in reduced heating costs. The dishwasher, in contrast, would actually consume more water and energy. This is not to say that no one should ever wash dishes. Rather, when considering how to invest time, money, and effort for the sake of environmental benefits, choosing the most effective actions, though not terribly difficult or daunting, requires a little bit of thought and homework.

Even if you were to adopt every rational energy conservation and recycling measure that you could find around your home and office, you might still feel discouraged because most everyone around you seems oblivious. Return to step one—lend them this book and the ASE and UCS books. I have not yet succeeded in convincing my sister-in-law to run the dishwasher only when it is full so as to make more efficient use of hot water, but I do intend to give her a copy of this book.

Remember, too, that only a few years ago very few cities had curbside recycling programs, but they are abundant now, thanks to the diligence of those promoting recycling. When parents do not recycle these days, they usually get a lecture from their kids, thanks to the efforts of teachers. Habits have changed regarding recycling, although there is

> "But now," says the Once-ler,
> "Now that you're here,
> the word of the Lorax seems perfectly clear.
> UNLESS someone like you
> cares a whole awful lot,
> nothing is going to get better.
> It's not."
> **—Dr. Seuss,** *The Lorax*

still plenty of room for further improvement. Even if the progress is incomplete and slow, take comfort in knowing that you are not too much of a hypocrite and that, for the most part, you are practicing what you preach.

Why do I say "not too much of a hypocrite" and "for the most part" practicing what you preach? A purist might argue that any use of fossil fuel is hypocritical if you truly believe in sustainable use of resources, because fossil fuels are irreplaceable. So to some degree, all of us who drive cars and heat our houses with electricity, gas, or oil are hypocrites. It does not bother me too much that I am a little bit of a hypocrite. Those who use only wood or only solar heating are still exceptional in our society, although the day when solar energy and wind power take off may be close at hand. In the meantime, those of us still using gas or oil must settle for what savings we can get from setting the thermostat a bit lower, installing insulation, and driving fuel-efficient cars. Setting the thermostat in winter at 66°F saves more energy and makes us less hypocritical than setting it at 68°F, and both are far better than 72°F. As President Jimmy Carter learned, however, people do not like to be preached to about their thermostat settings, and there is no such thing as a "politically correct" or an "environmentally correct" thermostat setting. Although lower is better during the heating season, and higher is better during the air conditioning season, set the thermostat at your level of tolerance, using your own commonsense cost-benefit analysis, including nonmonetary costs and benefits. More importantly, be sure you understand and can explain to others why the thermostat setting or any other form of energy conservation is important.

Insist that your elected representatives understand why you can't eat GNP.   Ask candidates for elective office a simple question: "Whence comes wealth?" If their answer does not include natural resources like soil, water, forests, and oceans, then they don't get it. Vote for someone else. If none of the candidates can give the right answer, then return to step one—lend them and their aides this book. A politician simply proclaiming that he or she is an environmentalist or is for both jobs and the environment is not enough. Those words are cheap. Our policymakers must understand the fundamental basis for our wealth and well-being, which includes clever use of technology and economics and prudent management of irreplaceable natural resources. If you get this far, lesson two would be to see how the politicians respond to the question of how discount rates should be chosen for governmental cost-benefit analyses. Do they understand the consequences of discounting the value of the environment of our children and grandchildren?

The decisions that governments make have profound effects on the environment, so it does make a difference who gets elected. A group of moderate Republicans narrowly managed to fend off the most egregious anti-environmental legislation proposed by the newly elected Republican congressional leadership in 1995. A few votes the other way would have led to the demise of the soil conservation program and to passage of legislation that would have reversed the progress made on clean air and clean water by earlier bipartisan consensus in Congress.

You can also ask for specific positions that the politicians have on the "top-down" approaches listed earlier in this chapter. Would the candidate for the U.S. Senate vote for ratifying the Kyoto treaty on global warming? Would the candidate stand up to grazing and mining interests and increase the current below-market fees paid for grazing and mining on public land? Does the candidate distinguish between support for family planning and abortion? What about the T-word? Does the candidate have the guts to vote in favor of consumption *taxes* on use of energy, carbon, water, and pesticides, provided that they are partially or fully offset by reduced taxes on income?

Environmental issues should not be and in many cases are not politically partisan. For example, Democratic Vice President Al Gore is a well-known advocate of actions to avert global warming, but one of his staunchest opponents on this issue, Senator Byrd from coal-rich West Virginia, is also a Democrat. Several conservative Republicans are opposing even the simplest actions to improve energy efficiency of refrigerators and other appliances because, they claim, they do not believe in global warming, and the energy conservation efforts might hurt the GNP. Nevertheless, equally conservative Republican Senator Connie Mack is trying to build a bipartisan consensus for tax credits for energy conservation. Not coincidentally, Senator

Mack represents the state of Florida, much of which is only a few feet above sea level and, hence, is very susceptible to damage from rising sea levels caused by global warming.

The cynical explanation of these patterns of politics is that immediate self-interest always rules. However, our common long-term interests really are at stake, and so we cannot afford to yield to the cynics. Rising sea level may not be of immediate concern to West Virginians and their senator, but their immediate self-interest is linked both to their coal mining jobs and their dependence on soil, water, air, forests, and climate for their everyday well-being. All of us, West Virginians and Floridians, Democrats and Republicans, need jobs to buy food, but we can't eat GNP.

Democracy does not end at the ballot box, and elected officials must continue to feel pressure. Most governments act responsibly only when pressured by their constituencies. The government of Brazil actually has had some very stringent environmental laws on its books for many years, such as requiring that ranchers leave trees on half of their land when they cut the Amazonian rain forests to create cattle pastures. The president of Brazil recently increased this percentage by decreeing that 80 percent of a landholder's property must now remain forested. There are also laws requiring public hearings and environmental impact studies for mining operations. But until very recently, these laws generally were not enforced, because the loggers, ranchers, and miners were the ones with enough money to wield influence in government. Now that democratic institutions are gradually returning to Brazil, a number of organizations representing a civil society, such as rural workers unions, civil rights groups, and nongovernmental scientific research organizations, have cropped up and are demanding a say, sometimes with success. People are empowered by information, and provided that at least some semblance of democratic institutions exist, informed people will demand responsiveness from their government.

The industrialized world is no different from Brazil in this respect. Our democracies are far from perfect, wealthy individuals and corporations have undue influence in government, and the public is nowhere nearly as well informed of environmental problems as it should be. Our few remaining old-growth rain forests in the Pacific Northwest are being logged, our soils are eroding, and we are pumping out groundwater faster than it is being replenished, as well as polluting the groundwater with agricultural and industrial chemicals. Keeping to an arcane nineteenth-century law designed for the Wild West, we allow mining companies to buy mining rights for a song and then to destroy the land surface and all the plants and animals on it while digging up the minerals. And we are addicted to wasteful use of coal, oil, and gas. Who are we to tell the Brazilians to save their rain forests? We must inform ourselves and our political leaders about our own problems and demand that we change our habits as well as change our governmental policies.

Figure out how you can best use your particular set of skills to contribute to helping change the way our society currently values and uses our resources. For me, my first major effort has been writing this book. As one who writes primarily for scientific journals read only by other scientists, I had to challenge myself to write in a completely dif-

> When we built Riverside, we called in the preeminent river conservation group for advice. Putting in silt barriers and doing thoughtful things with the water runoff was not an exceptional cost in our overall budget. It requires commitment and management time, but it's not unreasonable to expect developers to do the right thing. Shame on anybody who doesn't feel inclined to make the right decision. And that means everybody—developers, planners, businesspeople, environmentalists, CEOs, chambers of commerce, all of us.
> **—John Williams, Atlanta developer**

ferent style, which turned out to be much more difficult than writing for scientific journals. Each reader will have a different set of skills and will find different ways to challenge himself or herself to do something that will make a difference. It may not be easy. It might be suggesting a change in the workplace that would avoid waste of energy, paper, or building materials, or it might be a suggestion that would increase recycling or minimize production of garbage or pollution in the first place.

Some people are gifted at organizing fellow workers, neighbors, and friends into groups that take on projects. Teaching, both formally in the schools and informally in scout troops and fellowship groups, has a multiplier effect that will spark new interest and will start the process of training the next generation of teachers, workers, organizers, policymakers, scientists, and, don't forget, economists. If I did not hit on something that you think you can do, think of something else. Giving away money to environmental organizations or to green candidates running for

elected office is commendable, but giving money does not
let you off the hook for doing something else.

"Design Is the First Signal of Human Intention"

Trying to emit fewer pollutants or to produce less garbage
is commendable, but even better would be designing new
products and thinking of new ways of doing business and
carrying out our lives that do not emit pollutants to begin
with. That is the goal of a new office building that my
home institution, The Woods Hole Research Center, is designing. Rather than tearing down the existing house on
our new site, we are renovating it, thus minimizing the
refuse that must be hauled off to a dump. As in most office
buildings, we use lights, computers, FAX and photocopy
machines, coffee pots and a refrigerator. Working with a
group of architects and consultants, we are designing the
building to use natural lighting wherever possible not only
to save energy but also because humans prefer working
under natural lighting. Energy-efficient lighting, heating,
and air conditioning that can be controlled in each individual office will ensure that our workspace is comfortable—
we will prove that energy efficiency does not require sacri-
ficing comfort. We will achieve our goals without burning
any gas or oil in our building. A groundwater-source heat
pump will provide heating in the winter and cooling in the
summer. Much of our electrical energy will be generated
by a windmill on the site and by solar panels mounted on
the roof. Electricity drawn from the regional electrical grid
will help provide our demands on cloudy and windless
days, and we will send energy back to the grid, causing our

electric meter to run backwards, during windy days and sunny days.

As the wind and solar technologies develop and become more economically advantageous, we will expand this capacity and further reduce our dependence on the regional electrical grid. Our design is not locked into late 1990s technology but rather is designed to evolve as technological opportunities develop. In the meantime, all of our energy consumption habits will be scrutinized for the dual goals of energy efficiency and comfort of the workers. Our big clunky computer monitors, for example, will be replaced by flat screen monitors that use much less electricity, take up less desk space, and are easy on the eyes.

Our septic system will include a series of filter beds in which soil microorganisms convert urea and nitrate from our sewage into harmless nitrogen gas, thus providing a working example to our Cape Cod neighbors of an alternative to the common septic systems that are polluting the area's ponds and estuaries. The landscaping includes a grassy lawn big enough for picnics and for playing volleyball, but most of the area currently in lawn on this property will be converted to meadow or woodland that does not require pesticides, herbicides, and fertilizers. If we do use fertilizers on the lawn, any runoff from the fertilized area will pass through the downslope meadow and woodland, which will absorb the fertilizer runoff before it gets to the storm sewer in the street. Likewise, runoff from the parking area will be directed through a wooded area to permit infiltration into the soil and filtering of contaminants.

Many of these designs are common sense. Others have been suggested by experts. Every worker is being asked to contribute ideas about how to design our new home to be attractive, efficient, and comfortable. This interaction between the architects, energy consultants, and the common sense of the office inhabitants has been rich. Our goal is to use existing and affordable technologies that could be copied, in part or in whole, for other construction projects of office buildings and homes.

Another example of change underway is a worldwide alliance of 241 cities and counties called Cities for Climate Change, which is a program of the International Council for Local Environmental Initiatives (http://www.iclei.org) intended to help redesign the way that cities function. Impatient with national governments that have been slow to adopt effective policies to curb greenhouse gas emissions, these cities are voluntarily instituting steps to reduce emissions. By improving a solid waste recycling program and mass transit, Miami–Dade County in Florida has reduced greenhouse gas emissions by 1 million tons per year. The impacts of hurricanes where the land is only a few feet above sea level has particularly sensitized this southern Florida population to the effects of global warming, but other cities in the American West and Midwest, Canada, Germany, Hungary, India, Uganda, Vietnam, Israel, Italy, and Iran are also participating. An emphasis is being placed on the multiple benefits, such as reduced traffic and smog, that result from lowering greenhouse gas emissions. Better design of cities will improve their habitability as well as contribute to reducing local and global environmental problems.

At a time when many governments, particularly the United States Congress, are dragging their feet in implementing the top-down national governmental policies that are urgently needed, action by local governments and examples set by local institutions, homeowners, and citizens can push us in the right direction. Real change will occur when people actually change their habits, and you do not need to be an activist, organizer, or writer to set examples that challenge old habits. Just as my home institution is setting an example of office renovation, every homeowner, renter, office manager, or office worker can set examples, large and small, of how to reexamine old habits and adopt new ones that make sense. Simply setting examples of using resources responsibly will go a long way toward demonstrating to others that it can be done, that it is important to do, and that you value the resources that your children and grandchildren will inherit.

A SEA CHANGE

A change in habits goes hand in hand with changes in our way of thinking about how the world works. Our current system values the growth of gross national product over conservation of essential natural resources, and it falsely separates the economic system from the ecological system upon which our well-being depends. The economic pyramid must be placed within the ecological pyramid for either system to work well into the foreseeable, sustainable future. Futuristic technologies are not the most important ingredient for solving our environmental problems, although some of them will certainly help. Rather, we must change our way of thinking by acknowledging that economic and ecological systems are interdependent. This change in thinking will need to occur at all levels of society, from the homeowner installing an energy-efficient lightbulb to the World Bank economist approving a multimil-

lion-dollar forestry project or denying a government-subsi-
dized road-building project.

Many readers will find admonitions for being good,
energy-efficient citizens, like the list above, to be rather
unsatisfying. I often feel the same way until I remind myself that these types of actions are precisely the way that
widespread and profound changes in thinking take hold.
When the Berlin Wall was torn down and when the system
of apartheid in South Africa was dismantled, two ways of
thinking were quickly reversed that had seemed intransi-
gent only a few years earlier. No single individual or event
can be credited with those profound changes. Rather, at
some point, those economic and social systems became intolerable by the majority of the citizens of those countries
and by the rest of the world. People throughout the world
persistently opposed those forces of repression in both
small and large ways. A few people took active governmental and nongovernmental roles, both locally and internationally, but they were supported by the rest of us who
watched with great concern and who expressed, in our
own modest individual ways, our outrage over Eastern European communism and South African racism. Eventually,
these values prevailed over greed, power, corruption, ideology, and irrationality, although it is hard to put a finger
on exactly how this popular view overcame these seemingly immovable obstacles and actually caused a change.
Persistent, pervasive, popular pressure can apparently
eventually overwhelm strong adversaries.

We should also be outraged by our current system of discounting, devaluing, and degrading the environment, and
indeed, most of us already are. There are signs of hope, as
environmental issues have already moved from the politi-

cal fringe to the center. Television companies in Brazil have learned that their audience share increases when they add journalism about the environment in their news programming, apparently because Brazilians, like people from many other countries and cultures, are genuinely interested and concerned about the natural world around them. Even the World Bank, formerly the bastion of neoclassical economists looking for ways to increase global GNP, has a growing number of ecologists and ecological economists on its staff, and it now formally recognizes environmental effects as a consideration in its project review process.

On the other hand, the winning slogan in recent American presidential politics was not, "It's the Ecology, Stupid!" We have a long way to go to make the essential links between ecology and economics in the minds of policymakers, business leaders, voters, and consumers. Changing a way of thinking as deeply entrenched as neoclassical economics appears daunting, and I cannot put my finger on exactly how it will come about. Persistent, pervasive, popular pressure will surely be a part of it. A modest start is for you and me to make our case that you can't eat GNP.

Notes

The purpose of this section is to add further explanations where they are needed, to provide specific references to sources of my information, and to provide suggested reading for both the scientist and nonscientist interested in these topics.

CHAPTER 1

p. 3 Gross national product is known as Produto Nacional Bruto (PNB) in Portuguese and Spanish; Produit National Brut (PNB) in French; and Bruttosozialproduct (BSP) in German. Gross domestic product (GDP) is now used more commonly by economists than is GNP, but the later is used here because the public is more familiar with GNP. The GNP of a country includes the foreign earnings of its citizens and corporations, but excludes the earnings of foreign individuals and corporations from their activities within the country. In contrast, GDP includes all income earned within the country, regardless of the nationalities of those who made it, but excludes earnings of its citizens and corporations from activities conducted abroad. This distinction between GNP and GDP is not particularly germane to the topic of this book.

p. 6 The quotation of Arthur Burns was taken from a book by Thomas Prugh, Robert Costanza, John H. Cumberland, Herman Daly, Robert Goodland, and Richard B. Norgaard, *Natural Capital and Human Economic Sur-*

vival (Solomons, MD: International Society for Ecological Economics Press, 1995).

p. 7   The point about global warming having little direct effect on the economy because agriculture comprises only 3 percent of GNP was made by William Nordhaus in *Science* 253 (September 1991):1206. It should be noted, however, that Dr. Nordhaus was one of the leaders of the group of economists who issued a statement in 1997 acknowledging the risks of climate change and recommending that preventive steps are justified (see Chapter 5). In fairness to Dr. Nordhaus, I realize that he is one of the economists who probably does understand that you can't eat GNP. As this book explains, however, I disagree with many of the underlying assumptions of neoclassical economics, used by Dr. Nordhaus and other economists, which tend to undervalue the importance of a healthy ecological system as the basis for a healthy economic system.

Suggested further reading:

For a thorough and well-documented rebuttal of Rush Limbaugh, Gregg Easterbrook, and others espousing an anti-environmental and anti-science sentiment in the 1990s, I recommend Paul and Anne Ehrlich's book, *Betrayal of Science and Reason* (Washington, D.C.: Island Press, 1996). In this book, they analyze what was correct and incorrect in Paul Ehrlich's classic book, *The Population Bomb* (New York: Ballantine, 1968), as well as provide thoroughly referenced rebuttals to what they call "the brown backlash" that is currently trying to obstruct and reverse progress on sound environmental management.

CHAPTER 2

p. 15   The history of the Richland Creek tract of land, now known as the Carl Alwin Schenck Memorial Forest, was studied by John L Warren as a class project in 1972 at the Department of Forestry, North Carolina State University. I am indebted to Professor Charles B. Davey for finding and sharing with me Warren's report and for introducing me to forest soils, including our field trip to the Schenck Forest.

p. 24,   The quotations of Aldo Leopold are from *A Sand County Almanac* (New York: Ballantine Books, 1970). *(See also pp. 33, 41).*

p. 25   Data on the extent of soil erosion and land degradation and current rates of loss of agricultural land were drawn from three sources: C. J. Barrow, *Land Degradation* (Cambridge: Cambridge University Press, 1991); David Pimental et al., "Environmental and Economic Costs of Soil Erosion and Conservation Benefits," *Science* 267 (1995):1117–1123; and Gretchen Daily, "Restoring Value to the World's Degraded Lands," *Science* 269 (1995):350–354.

Another well-balanced discussion of the strengths and weaknesses of these estimates of land degradation can be found in a report by the Stockholm Environment Institute (SEI), *Global Land and Food in the 21st Century,* POLESTAR Series Report no. 5, 1995. SEI, Box 2142, S–103 14 Stockholm, Sweden.

p. 26   The late Dr. Hans Jenny, a soil scientist at the University of California at Berkeley, wrote a historical account of how E. W. Hilgard identified the problems of irrigation of California farmland in the late nineteenth century: *E. W. Hilgard and the Birth of Modern Soil Science,* Industrie Grafiche V. Lischi & Figli, Pisa, Italy, 1961. Hilgard, who has been called the American father of soil science and for whom the soil science building at Berkeley is named, did not know about selenium, but he did recognize the process by which excessive irrigation of arid land would lead to accumulation of several kinds of salts, rendering the farmland unusable for agriculture.

Suggested further reading:
An excellent introduction on the importance of soil, a historical account of how humans have used and abused soil, and an expression of inspirational reverence toward the role of soil in civilization can be found in Daniel Hillel, *Out of the Earth* (Berkeley: University of California Press, 1991). My examples of lessons learned from the ancient Greeks and Romans and of both ancient and modern water conservation in the Middle East are drawn from Hillel's book.

A history of the development of agricultural technology can be found in L. T. Evans, "The Natural History of Crop Yield," *American Scientist* 68 (1980): 388–397; and in Rosamond Naylor, "Energy and Resource Constraints on Intensive Agricultural Production," *Annual Review of Energy and the Environment* 21 (1996):99–123.

CHAPTER 3

p. 38 The cost-benefit analysis of auto safety features appeared in the April 1996, issue of *Consumer Reports.*

p. 41 A discussion of the spread of tropical diseases expected with global warming is reported by Richard Stone, *Science* 267 (February 17, 1995):957–958; by Jonathan A. Pratz et al., *Journal of the American Medical Association* 275 (January 17, 1996):217–223; and by Paul Epstein, *Science* 285 (July 16, 1999):347–348. The debate over the value of a human life surfaced during evaluation of a report on the economic and social implications of global warming by one of the working groups of the Intergovernmental Panel on Climate Change. The story is covered in Ehsan Masood and Ayalala Ochert, "UN Climate Change Report Turns Up the Heat," *Nature* 378 (November 9, 1995):119.

p. 44   An account of New York City's investment in its forested watersheds can be found in Graciela Chichilnisky and Geoffrey Heal, *Nature* 391 (February 12, 1998):629–630.

p. 46   The assessment of the economic value of ecosystem services was reported by Robert Costanza and colleagues in "The Value of the World's Ecosystem Services and Natural Capital," *Nature* 387 (May 15, 1997):253–260. A follow-up story, with responses from several neoclassical and ecological economists can be found in Ehsan Masood and Laura Garwin, "Costing the Earth: When Ecology Meets Economics," *Nature* 39 5(October 1, 1998):428–434.

p. 50   The beneficial effects of fruits and vegetables in the diet have been explained at length by my mother (personal communication), as well as in the report, Walter C. Willett, "Diet and Health: What Should We Eat?" *Science* 264 (1994):532–537.

p. 51   An interesting series of publications demonstrate how different interpretations can be reached about the risks of the pesticide Alar, while using the same sets of data. In chronological order: Leslie Roberts, "Alar: The Numbers Game" *Science* 243 (1989):1430; Eliot Marshall, "A Is for Apple, Alar, and . . . Alarmist?" *Science* 254 (1991):20–22; Adam Finkel and Lawric Mott, "Alar: The Aftermath" *Science* 255 (1992):664–665.

Suggested further reading:

A good introductory book on the new discipline of ecological economics is Robert Costanza, John Cumberland, Herman Daly, Robert Goodland, and Richard Norgaard, *An Introduction to Ecological Economics* (Boca Raton, FL: St. Lucie Press, 1997). Another excellent book by many of the same authors is Thomas Prugh, Robert Costanza, John H. Cumberland, Herman Daly, Robert Goodland, and Richard B. Norgaard, *Natural*

Capital and Human Economic Survival, (Solomons, MD: International Society for Ecological Economics Press, 1995).

An analysis of the many services that natural ecosystems provide humanity is the focus of a book compiled by scientific scholars supported by the Pew Foundation: G. Daily, ed., *Nature's Services: Societal Dependence on Natural Ecosystems* (Washington, D.C.: Island Press, 1997).

Chapter 4

p. 72   The economic analysis that attempts to justify delaying actions to avoid global warming was published by T.M.L. Wigley et al., "Economic and Environmental Choices in the Stabilization of Atmospheric CO_2 Concentrations," *Nature* 379 (January 18, 1996):240–243. Dr. Wigley has since stated that this paper has been widely misinterpreted and that he did not intend it to be a justification for delays of energy conservation actions that would help to avoid global warming (*Nature* 383 [October 24, 1996]:657). Nevertheless, Wigley later published another paper, "Implications of Recent CO_2 Emission-Limitation Proposals for Stabilization of Atmospheric Concentrations," *Nature* 390 (November 20, 1997):267–270, in which he again uses the same assumptions of discounting the future and of technological super-optimism (Custer's folly) to suggest that there is no urgency to actions to avert global warming. His arguments were rebutted by Kilaparti Ramakrishna, "The Great Debate on CO_2 Emissions," in the same issue (pp. 227–228).

Suggested further reading:
The topic of discounting is explained further in John Gowdy and Sabine O'Hare, *Economic Theory for Envi-*

ronmentalists (Delray Beach, FL: St. Lucie Press, 1995). A more in-depth analysis of how discounting relates to sustainability is given by Richard B. Norgaard and Richard B. Howarth in "Sustainability and Discounting the Future," in *Ecological Economics*, ed. Robert Costanza (New York: Columbia University Press, 1991), pp. 88–101. David Pearce and coauthors also devote a chapter to "Discounting the Future" in their book, *Blueprint for a Green Economy* (London: Earthscan Publications, 1989).

CHAPTER 5

p. 85   There have been several reports of the Intergovernmental Panel on Climate Change. The most recent comprehensive ones include the following: J. T. Houghton et al., eds., *Climate Change 1995—The Science of Climate Change: Contribution of Working Group I to the Second Assessment Report of the Intergovernmental Panel on Climate Change* (Cambridge: Cambridge University Press, 1996); R. T. Watson et al., eds., *Climate Change 1995—Impacts, Adaptations, and Mitigation of Climate Change: Scientific-Technical Analyses: Contribution of Working Group II to the Second Assessment Report of the Intergovernmental Panel on Climate Change* (Cambridge: Cambridge University Press, 1996); J. Bruce et al., eds., *Climate Change 1995—Economic and Social Dimensions of Climate Change: Contribution of Working Group III to the Second Assessment Report of the Intergovernmental Panel on Climate Change* (Cambridge: Cambridge University Press, 1996). The IPCC reports contain scientific reviews of the current state of knowledge and uncertainties about global warming as well as executive summaries that are written for nonscientists. The Third Assessment Report of the IPCC is due out by the end of the year 2000.

p. 87 The record high temperatures of the late 1980s have resumed in the late 1990s, after a brief interruption caused by the cooling effect of aerosols emitted into the atmosphere by the volcanic eruption of Mt. Pinatubo in the Philippines in 1991 (see M. Patrick McCormick et al., "Atmospheric Effects of the Mt. Pinatubo Eruption," *Nature* 373 (February 2, 1995):399–404. The temporary cooling caused by the Pinatubo eruption actually strengthens our predictions about global warming, because the climate system behaved just as the global climate models predicted; that is, the warming resumed as soon as the dust settled from the Pinatubo eruption. A chronological progression of climatologist James Hansen's mostly correct predictions about global warming and the effects of Mt. Pinatubo can be traced through the following series of news articles by Richard Kerr: "Hansen vs. the World on the Greenhouse Threat," *Science* 244 (June 2, 1989):1041–1043; "Global Temperature Hits Record Again," *Science* 251 (January 18, 1991):274; "1991: Warmth, Chill May Follow," *Science* 255 (January 17, 1992):281; "Pinatubo Global Cooling on Target," *Science* 259 (January 29, 1993):594; "1995 the Warmest Year? Yes and No," *Science* 271 (January 12, 1996):137–138; and "The Hottest Year, by a Hair," *Science* 279 (January 16, 1998):315–316.

Degrees of certainty and uncertainty, from "virtually certain facts" to "probable projections" about the effects of global warming are described by J. D. Mahlman in "Uncertainties in Projections of Human-Caused Climate Warming," *Science* 278 (November 21, 1997):1416–1417. For a scientific account of increased variability in rain and snow events, see the article by A. A. Tsonis, "Widespread Increases in Low-Frequency Variability of Precipitation Over the Past Century," *Nature* 382 (August 22, 1996):700–702.

A thorough listing of the mounting evidence of global warming is beyond the scope of these notes, but what follows is a partial list to illustrate how the effects are being observed in the Arctic tundra, Walter Oechel et al., "Recent Change of Tundra Ecosystems from a Net Carbon Dioxide Sink to a Source," *Nature* 361 (February 11, 1993):520–523; lakes within boreal forests, D. Schindler et al., "Effects of Climatic Warming on Lakes of the Central Boreal Forest," *Science* 250 (November 16, 1990):967–970; glaciers, J. Oerlemans and J. Fortuin, "Sensitivity of Glaciers and Small Ice Caps to Greenhouse Warming," *Science* 258 (October 2, 1992):115–120; polar sea ice, C. Doake et al., "Breakup and Conditions for Stability of the Northern Larren Ice Shelf, Antarctica," *Nature* 391 (February 19, 1998):778–780, Michael Oppenheimer, "Global Warming and the Stability of the West Antarctic Ice Sheet," *Nature* 393 (May 28, 1998):325–332, and Richard A. Kerr, "Will the Arctic Ocean Lose All Its Ice," Science 286 (December 3, 1999):1828; mountain plants, Georg Grabherr et al., "Climate Effects on Mountain Plants," *Nature* 369 (June 9, 1994):448; grasslands, Richard Alward et al., "Grassland Vegetation Changes and Nocturnal Global Warming," *Science* 283 (January 8, 1999):229–231; birds, Humphrey Crick and Timothy Sparks, "Climate Change Related to Egg-Laying Trends," *Nature* 399 (June 3, 1999):423–424; and coral reefs, Elizabeth Pennisi, "New Threat Seen from Carbon Dioxide," *Science* 279 (February 13, 1998):989; Peter Pock, "Global Warming 'Could Kill Most Coral Reefs by 2100'" Nature 400 (July 8, 1999):98.

p. 88   The full statement signed by more than 2500 scientists can be found at: http://www.ozone.org. The full statement signed by more than 2000 economists can be obtained from Redefining Progress, One Kearny St., Third

Floor, San Francisco, CA 94108 On-line posting: info@rprogress.org.

p. 90   The following scientific papers and reports discuss the rate at which trees can be expected to migrate relative to past and future warming: Leslie Roberts, "How Fast Can Trees Migrate?" *Science* 243 (February 10, 1989):735–737; Richard Kerr, "Greenhouse Report Foresees Growing Global Stress," *Science* 270 (November 3, 1995):731; Camille Parmesan, "Climate and Species' Range," *Nature* 382 (August 29, 1996):765–766; Gordon C. Jacoby et al., "Mongolian Tree Rings and 20th-Century Warming," *Science* 273 (August 9, 1996):771–773; and J. T. Overpeck et al., "Potential Magnitude of Future Vegetation Change in Eastern North America: Comparison with the Past," *Science* 254 (November 1, 1991):692–695.

p. 95   A pair of papers that provide contrasting conclusions from cost-benefit analyses by economists regarding air pollution control in the Los Angeles area were published by Alan Krupnick and Paul Portney in "Controlling Urban Air Pollution: A Benefit-Cost Analysis, *Science* 252 (April 26, 1991):522–527; and by Jane Hall et al. in "Valuing the Health Benefits of Clean Air," *Science* 255 (February 14, 1992):812–817. Although they use very similar data as a base for their analyses, they use different sets of assumptions, which produce different estimates of whether the monetary value of the human health benefits of improved air quality exceeds the monetary costs of pollution control.

p 97   One example of documentation of a decline in lead in the environment following the ban on leaded gasoline is in a report by Rosman et al., "Isotopic Evidence for the Source of Lead in Greenland Snows Since the late 1960s," *Nature* 362 (March 25, 1993):333–335. Excellent reviews of the problem of depletion of the ozone layer in the stratosphere and how industry first opposed and later supported restrictions of CFCs can be found in papers by Michael Prather et al., "The Ozone Layer: The Road Not Taken," *Nature* 381 (June 13, 1996):551–554; and Nobel Prize winner F. Sherwood Rowland, "President's Lecture: The Need for Scientific Communication with the Public," *Science* 260 (June 11, 1993):1571–1576. The low costs of reducing acid rain by using tradable pollution permits is chronicled by Richard Kerr in "Acid Rain Control: Success on the Cheap," *Science* 282 (November 6, 1998):1024–1027.

p. 102 The "drop-in-the-bucket" but nevertheless a "step-in-the-right-direction" interpretation of the climate change convention completed in Kyoto, Japan, in 1997 is discussed by Bert Bolin, in "The Kyoto Negotiations on Climate Change: A Scientific Perspective," *Science* 279 (January 16, 1998):330–331.

p. 104 When we will run out of cheap oil is discussed by Criag Hatfield in "Oil Back on the Global Agenda," *Nature* 387 (May 8, 1997):121, and in a subsequent correspondence in *Nature* 388 (August 14, 1997):618. The topic is also reviewed by Richard Kerr in "The Next Oil Crisis Looms Large—and Perhaps Close," *Science* 281 (August 21, 1998):1128–1131.

p. 107 The role of aerosols in climate is reviewed by K. Hassemann in "Are We Seeing Global Warming?" *Science* 276 (May 9, 1997):914–915. The importance of dust from increased wind erosion of soils is reviewed by Meinrat O. Andreae in "Raising Dust in the Greenhouse," *Nature* 380 (April 4, 1996):389–390.

p. 110 The objections to the IPCC report by the coal and oil industry are covered in Ehsan Masood, "Head of Climate Group Rejects Claims of Political Influence," *Nature* 381 (June 6, 1996):455, and Peter Weiss, "Industry Group Assails Climate Chapter," *Science* 272 (June 21,

1996):1734. John Houghton responded in a letter to *Nature* 382 (August 22, 1996):665. The withdrawal of two companies from the industry-supported Global Climate Coalition was reported in Ehsan Masood, "Companies Cool to Tactics of Global Warming Lobby," *Nature* 383 (October 10, 1996):470. The financial support provided by the oil and coal industry to the handful of scientists actively opposing the IPCC consensus report has been documented in "The Heat Is On" by Ross Gelbspan, *Harper's Magazine,* (December 1995) and in a book by the same title published by Addison-Wesley Publishing Company in 1997.

p. 113   The winter blizzard of 1996 was a cover story in *Newsweek Magazine* (January 22, 1996), where the quotation by Stephen Schneider appears. The storm was also covered in the *New York Times* (January 14, 1996, "Blame Global Warming for the Blizzard," by William K. Stevens).

p.116   Report of the insurance industry's response to global warming can be found in Alison Abbott, "Insurance Company to Back Out of Some Climate-Linked Risks," *Nature* 372 (November 17, 1994):212–213, and "Climate Change Disasters Looming, Warns Insurer," *Nature* 394 (July 30, 1998):412.

p. 117   Growing evidence of the linkage between global warming and the frequency and severity of El Niño events is reported in Michael McPhaden, "The Child Prodigy of 1997–98," *Nature* 398 (April 15, 1999):559–562, and presented in a scientific paper by Timmermann et al. in *Nature* 398 (April 22, 1999):694–697.

Suggested further reading:
J. T. Houghton et al., eds., "Summary for Policymakers," in *Climate Change 1995—The Science of Climate Change: Contribution of Working Group I to the Sec-*

ond Assessment Report of the Intergovernmental Panel on Climate Change (Cambridge: Cambridge University Press, 1996). John Firor, *The Changing Atmosphere: A Global Challenge* (Cambridge: Yale University Press, 1992). Stephen H. Schneider, *Laboratory Earth: The Planetary Gamble We Can't Afford to Lose* (London: Weidenfeld and Nicolson, 1996).

CHAPTER 6

p. 123   Gregg Easterbrook's book, *A Moment on the Earth* (New York: Viking, 1995), is an unfortunate example of misplaced complacency about environmental problems.

p. 125   The quotation of Daniel Hillel is from his book, *Out of the Earth* (Berkeley: University of California Press, 1991).

p. 126   The full report of the National Research Council on the transition toward sustainability is published as a book: *Our Common Journey* (Washington, D.C.: National Academy Press, 1999).

p. 126   Data on use and abuse of groundwater throughout the world can be found in the following: P. H. Gleick, ed., *Water in Crisis* (New York: Oxford University Press, 1993); and S. Postel et al., "Human Appropriation of Renewable Fresh Water," *Science* 271 (1996):785–788.

p. 128   Arsenic in well water in India was reported in *Science* 274 (October 11, 1996):174–175.

p. 129   Discussions of overpumping of the Ogallala aquifer and diversion of water from the Aral Sea can be found in the book by Hillel, cited above.

p. 132   Garret Hardin's paper titled "Tragedy of the Commons" was published in *Science* 162 (December 13, 1968):1243–1248. A challenge to the "tragedy of the commons" concept in fisheries can be found in the spring 1995 edition of *The Ecologist* (25:46–73).

p. 133   The quotations of Aldo Leopold are from *A Sand County Almanac* (New York: Ballantine Books, 1970).

p. 135   The destruction of the Aral Sea is also covered by Hillel (see above) and by Richard Stone, "Coming to Grips with the Aral Sea's Grim Legacy." *Science* 284 (April 2, 1999):30–33. Vice President Al Gore also vividly describes his experience visiting the Aral Sea in his book *Earth in the Balance* (Boston: Houghton Mifflin, 1992).

p. 136   A discussion of designing effective use of resources, including the design of eventual recycling when a product is originally manufactured, such as the example given for carpet, is offered by William McDonough and Michael Braungart in their October 1998 article in *Atlantic Monthly,* titled "The NEXT Industrial Revolution."

Suggested further reading:

In his recent book, *The Ostrich Factor* (New York: Oxford University Press, 1999), Garrett Hardin reexamines the writings of Malthus.

A readily accessible account of the groundwater crisis is given in Sandra Postel, *Dividing the Waters,* Worldwatch Paper 132 (Washington, DC: Worldwatch Institute, 1996).

CHAPTER 7

p. 142   The quotation from Gro Brundtland is from the lead editorial in the July 25, 1997, issue of *Science* (277:457).

p. 143   For another popular account of Cássio Pereira's work, along with his advisor, Daniel Nepstad, see Gabrielle Walker, "Slash and Grow," *American Scientist,* (September 21, 1996), 28–33.

pp. 146   Norman Myers's discussion of the "shifted cultivator" appears in his paper "Tropical Deforestation: Population, Poverty, and Biodiversity," in *The Economies and*

Ecology of Biodiversity Decline: The Forces Driving Global Change, ed. T. M. Swanson (Cambridge: Cambridge University Press, 1995), pp. 111–122.

p. 154   Herman E. Daly has published numerous works on macroeconomics and sustainability, including his contribution to *Natural Capital and Human Economic Survival* and *An Introduction to Ecological Economics*, cited in the notes for Chapters 1 and 3. He also wrote *Beyond Growth: The Economics of Sustainable Development* (Boston: Beacon Press, 1996).

p. 155   Kenneth Boulding used the metaphors of a cowboy economy and a spaceship economy in his classic 1966 article "The Economics of the Coming Spaceship Earth." This essay has been reprinted on pages 297–309 in a collection of classic essays, *Valuing the Earth*, edited by Herman E. Daly and Kenneth N. Townsend, The MIT Press, Cambridge, Massachusetts, 1993. Herman Daly reviews Boulding's ideas in his chapter "Elements of Environmental Macroeconomics," in *Ecological Economics*, ed. Robert Costanza (New York: Columbia University Press, 1991), pp. 32–46.

p. 156   "The Ballad of Ecological Awareness" appears on page 109 in a book by Thomas Prugh, Robert Costanza, John H. Cumberland, Herman Daly, Robert Goodland, and Richard B. Norgaard, *Natural Capital and Human Economic Survival* (Solomons, MD: International Society for Ecological Economics Press, 1995).

p. 157   Garrett Hardin's book *The Ostrich Factor* is cited in the notes for Chapter 6.

p. 158   Quotations of Edward O. Wilson are from his book *Biophilia* (Cambridge: Harvard University Press, 1984).

Suggested further reading:

I find the work on deforestation of the Amazon rain forest by my colleagues Dan Nepstad and Chris Uhl to be particularly insightful about how social, economic, and ecological factors are intertwined. Here are a few samples of their work: C. Uhl et al., "An Ecosystem Perspective on Threats to Biodiversity in Eastern Amazonia, Pará State," in *Perspectives on Biodiversity: Case Studies of Genetic Resource Conservation and Development,* ed. C. S. Potter et al. (Washington, DC: American Association for the Advancement of Science, 1993); Christopher Uhl et al., "Natural Resource Management in the Brazilian Amazon," *BioScience* 47 (1997):160–168; A. Verissimo et al., "Extraction of a High-Value Natural Resource in Amazonia: The Case of Mahogany," *Forest Ecology and Management* 72 (1995):39–60; Dan Nepstad et al., "Flames in the Rain Forest: Origins, Impacts, and Alternatives to Amazonian Fire," Pilot Program to Conserve the Brazilian Rain Forest, Brasília, 1999 (available from the Woods Hole Research Center, P.O. Box 296, Woods Hole MA 02543); and Dan Nepstad et al., "Recuperation of a Degraded Amazonian Landscape: Forest Recovery and Agricultural Restoration," *Ambio* 20 (1991): 248–255.

A comprehensive assessment of the status of the world's forests, the multiple cultural and economic demands upon them, and suggestions for how to protect them can be found in the report of the World Commission of Forests and Sustainable Development, *Our Forests, Our Future* (Cambridge: Cambridge University Press, 1999).

CHAPTER 8

p. 162   The term *biodiversity* became commonly used in the fields of biology and ecology after Harvard entomologist E. O. Wilson published his book *BioDiversity* in 1988 (Washington, DC: National Academy Press).

p. 163   The estimates of deforestation and reforestation are for the 1981–1992 period by the United Nations Food and

Agriculture Organization as reported in *Forest Resources Assessment 1990, Tropical Countries.*

p. 167 I use, with apologies, some license in describing Paul Ehrlich's metaphor about the rivets on an airplane wing resembling species on the earth. The original metaphor can be found in his 1981 book with Anne Ehrlich, *Extinction: The Causes and Consequences of the Disappearance of Species* (New York: Random House).

p. 167 The quotations of Aldo Leopold are from *A Sand County Almanac* (New York: Ballantine Books, 1970).

p. 168 The estimates of numbers of species that exist, that have gone extinct, and that are at risk of extinction come from articles by Fred Powledge, "Biodiversity at the Crossroads," *Bioscience* 48 (May 1998):347–352; Fraser Smith et al., "Estimating Extinction Rates," *Nature* 364 (August 5, 1993):494–496; and Constance Holden, "Red Alert for Plants," *Science* 280 (April 17, 1998):385.

p. 169 The debate between Norman Myers and Julian L. Simon was published in *Scarcity or Abundance* (New York: W. W. Norton, 1994). In fairness to Myers, I should add that in a recent book review written by him (*Bioscience* 46:717–719, October 1996), he makes the same point that I make here, that the diversity being lost within species is probably more important than the number of species going extinct. He also emphasizes, as I have here, that saving species in parks and preserves is not sufficient.

p. 174 The case of the snail going extinct was reported in *Science* 282:215 (October 9, 1998):215.

p. 179 The quotation of Daniel Nepstad is from unpublished documents of the Woods Hole Research Center.

p. 182 An example of how logging in the Amazon Basin is often like mining can be found in A. Verissimo et al., "Logging Impacts and Prospects for Sustainable Forest Management in an Old Amazonian Frontier: The Case of

Paragominas," *Forest Ecology and Management* 55
(1992):169–199; and in Christopher Uhl et al., "Natural
Resource Management in the Brazilian Amazon," *Bio-
Science* 47 (1997):160–168.

Suggested further reading:
Two excellent articles on the biodiversity topic are Sir
Robert M. May, "How Many Species Are There on
Earth?" *Science* 241 (September 16, 1998):1441–1449;)
and Stuart L. Pimm et al., "The Future of Biodiversity,"
Science 269 (July 21, 1995):347–350.

Several references to Chris Uhl's work in the Amazon
have already been made in the notes for Chapters 7 and
8. For a more global view, *Saving the Forests: What Will
It Take?* Worldwatch Paper #117 (Washington, DC:
Worldwatch Institute, 1993) provides a concise
overview of the issue.

A more in-depth and up-to-date perspective, but one
that is still highly accessible for the nonscientist, is the
recent report of the World Commission of Forests and
Sustainable Development: *Our Forests, Our Future*
(Cambridge: Cambridge University Press, 1999).

CHAPTER 9

p. 188 A thorough analysis of the history and prospects for human population growth is provided by Joel E. Cohen in
his book *How Many People Can the Earth Support?*
(New York: W. W. Norton, 1995). Cohen also explores
the many social and economic reasons for the demographic transition from high birth and death rates to low
birth and death rates in many cultures throughout the
world.

p. 189 See Garrett Hardin, *The Ostrich Factor* (New York: Oxford University Press, 1999).

p. 190 The data on education and sanitation are from a UNICEF report, "Promise and Progress: Achieving Goals for Children" (New York: United Nations, 1996), also available at www.unicef.org/information/mdg, and from Constance Holden, "Global Water Scarce," *Science* 262 (November 12, 1993):987.

p. 195 The quotation of Theodore Panayotou is from his book *Green Markets* (San Francisco: Institute for Contemporary Studies Press, 1993).

p. 195 An excellent review of the collapse of major fisheries and the many possible causes and solutions is Simon Fairlie et al., "The Politics of Fishing," *Ecologist* 25 (Spring 1995):46–73.

p. 199 The comparison of the costs of governmental subsidies that create perverse incentives for environmental degradation versus the costs of a comprehensive conservation program was made by Alexander James et al., "Balancing the Earth's Accounts," *Nature* 401 (September 23, 1999):323–324. The lead editorial in the November 25, 1999, issue of *Nature* addresses the need to include environmental concerns by the World Trade Organization.

p. 202 The web site (http://www.ase.org) of the Alliance to Save Energy has a great deal of information, including a recommended home energy audit. Their literature can also be obtained by writing to 1200 18th St. NW, Suite 900, Washington, D.C., 20036, or calling 202–857–0666.

p. 203 The book sponsored by the Union of Concerned Scientists (UCS) is Michael Brower and Warren Leon, *The Consumer's Guide to Effective Environmental Choices* (New York: Three Rivers Press, 1999). The UCS can also be contacted at 2 Brattle Square, Cambridge, MA 02238–9105; 617–547–5552. Some of the same information can be obtained directly from their web site: http://www.ucsusa.org.

p. 205   Dr. Seuss, *The Lorax* (New York: Random House, 1971).

p. 210   The quotation of John Williams is from the August 1999 edition of the United Airlines magazine, *Hemispheres.*

p. 211   William McDonough is the lead architect of the Woods Hole Research Center's new building, as well as the author of the quotation used for this subheading. More information on our building's design can be found at: http://www.whrc.org.

p. 213   The Cities for Climate Change program was reported in the May 5, 1998, edition of *EOS, Transactions, American Geophysical Union.*

Index

About the Author

Dr. Eric A. Davidson is a senior scientist at the Woods Hole Research Center. He holds a bachelor's degree in biology from Oberlin College and a Ph.D. in forestry from North Carolina State University. His current research focuses on how soil properties and processes change when forests are converted to agricultural land and when agricultural land is allowed to revert back to forest. About half of his time is spent studying rain forests and cattle pastures in the Brazilian Amazon, and the other half is spent in the regrowing forests of New England.

Prior to moving to Woods Hole in 1991, he completed postdoctoral research fellowships in soil microbiology at the University of California, Berkeley, and in biogeochemistry at the NASA Ames Research Center, Moffett Field, California. He is an associate editor of the *Soil Science Society of America Journal,* and he has published more than fifty peer-reviewed scientific papers in journals that specialize in ecology, biogeochemistry, and soil science.

Eric and his then newlywed wife, Jean Talbert, served as Peace Corps volunteers from 1979 to 1981, working on public health projects in Zaire (now the Democratic Republic of the Congo). Their son, Bryce, was born in 1993.

George M. Woodwell, PhD., internationally known ecologist, is President and founder of the Woods Hole Research Center, and founding trustee of the Natural Resources Defense Council, the Environmental Defense Fund, and World Resources Institute.